T0172925

Fading and Interference Mitigation in Wireless Communications

OTHER COMMUNICATIONS BOOKS FROM AUERBACH

Fading and Interference Mitigation in Wireless Communications

Stefan R. Panić

Mihajlo Stefanović • Jelena Anastasov • Petar Spalević

CRC Press
Taylor & Francis Group
Boca Raton   London   New York

CRC Press is an imprint of the
Taylor & Francis Group, an **informa** business

CRC Press
Taylor & Francis Group
6000 Broken Sound Parkway NW, Suite 300
Boca Raton, FL 33487-2742

Printed on acid-free paper
Version Date: 20131023

International Standard Book Number-13: 978-1-4665-0841-5 (Hardback)

Visit the Taylor & Francis Web site at
http://www.taylorandfrancis.com

and the CRC Press Web site at
http://www.crcpress.com

Contents

Preface

Professional bodies of note for electrical engineers, such as IET and IEEE, have more than 500,000 members worldwide and hold more than 3,000 conferences annually. Many of the members of these bodies are interested in advanced digital telecommunications and signal processing. According to latest statistical figures, the United States produces 50,000 electrical engineers in a year as against 70,000 by India and 100,000 by China. Digital telecommunications and signal processing are mandatory subjects at most of the faculties and colleges in these countries. Considering these facts, this book could present itself as a useful literature for introducing some basic principles of digital communications and its mathematical formalization and also for staying updated with the recent developments in the performance analysis of space diversity reception over fading channels in the presence of cochannel interference (CCI).

Researchers have been studying problems associated with these domains for more than 40 years, and system engineers have used the theoretical and numerical results reported in the literature to guide the design of their systems. In recent years, applications have become increasingly sophisticated, thereby requiring more complex models and improved diversity techniques along with the need to provide performance analyses of the scenarios studied.

The primary goal of this book is to present a unified method to provide a set of tools that will allow the system designer to compute the performance of digital communication systems characterized by a variety of modulation and detection types and channel models.

The underlying concept of this work is to present an accurate performance evaluation of the proposed communication scenarios along with providing insight into the dependence of these performances on key system parameters.

The major topics covered are multichannel reception in various fading environments, influence of cochannel interference, and macrodiversity reception when channels are simultaneously affected by various types of fading and shadowing. The contents of this book are relevant not only for the wireless applications that are rapidly filling our technology journals but also for a host of other applications involving satellite, terrestrial, and maritime communications. Analysis applicability and subject actuality are very high.

The book is organized in such a manner as to guide the reader through a step-by-step process on the basics of performance analysis of digital communication receivers, providing them the necessary concepts of digital communication system design. This could serve as a time-saving reference as the readers could find, at one place, what is generally covered in a range of standard digital communication and signal processing courses, thus extending their knowledge beyond those courses. As a large collection of delivered system performances are available, this book will allow researchers and system designers to perform trade-off studies among the various communication type/fading channel combinations, allowing them to make an optimum choice.

This work could be treated as an extension and addendum to the results provided in the book written by Simon and Alouini [1] and to the results of the studies conducted by various well-known researchers in this field, such as G. Karagiannidis and M. Yacoub.

We would like to thank Rich O'Hanley for encouraging us to publish this book.

We would also like to thank Professor Dr. Sinisa Ilic and Professor Dr. Predrag Lekic for their help in preparing the manuscript.

Reference

1. Gradshteyn, I. and Ryzhik, I. (1980). *Tables of Integrals, Series, and Products*. Academic Press, New York.

Nomenclature

NOTATION	DESCRIPTION OR NAME OF FUNCTION AND REFERENCE CITATION FOR ITS DEFINITION
ABER	Average bit error probability
AFD	Average fade duration
AM	Amplitude modulation
AoF	Amount of fading
ASEP	Average symbol error probability
ASER	Average symbol error rate
ASK	Amplitude-shift-keying
AWGN	Additive white Gaussian noise
$B_x(p, q)$	Incomplete beta function [1, Eq. (8.391)]
BDPSK	Binary differential phase-shift keying
BEP	Bit error probability
BER	Bit error rate
BFSK	Binary frequency-shift keying
BPSK	Binary phase-shift keying
BS	Base station
CCI	Cochannel interference
CDF	Cumulative distribution function
CDMA	Code-division multiple access
CFSK	Coherent frequency-shift keying
CIFR	Channel inversion with fixed rate
$Cov(X, Y)$	Covariance operator
CPFSK	Continuous-phase frequency-shift keying
CPSK	Coherent phase-shift keying

NOTATION	DESCRIPTION OR NAME OF FUNCTION AND REFERENCE CITATION FOR ITS DEFINITION
CSI	Channel state information
DPSK	Differential phase-shift keying
DS-CDMA	Direct-sequence code-division multiple access
EGC	Equal gain combining
$erfc(z)$	Complementary error function [1, Eq. (8.250.4)]
$E(X)$	Mathematical expectation operator
$_1F_1(a, b; z)$	Kummer confluent hypergeometric function [1, Eq. (9.210.1)]
$_2F_1(a, b; c; z)$	Gaussian hypergeometric function [1, Eq. (9.14.2)]
$_pF_q(a_1, a_2, \ldots, a_p; b_1, b_2, \ldots, b_p, z)$	Generalized hypergeometric function [1, Eq. (9.14.1)]
FSK	Frequency-shift keying
$G_{p,q}^{m,n}\left(x\left\vert\begin{array}{c}a_1 \ldots a_p \\ b_1 \ldots b_q\end{array}\right.\right)$	Meijer's G-function [1, Eq. (9.301)]
GSC	Generalized selection combining
GSM	Global system for mobile communications
i.i.d.	Independent, identically distributed
ISI	Intersymbol interference
J	Jacobian transformation
JPDF	Joint probability density function
K	Rician fading parameter
LCR	Level crossing rate
LN	Lognormal
$L_n^k(x)$	Laguerre polynomial (generalized) [1, Eq. (8.970)]
LOS	Line of sight
M-AM	Multiple amplitude modulation
M-ASK	Multiple amplitude-shift keying
MC-CDMA	Multicarrier code-division multiple access
M-DPSK	Multiple differential phase-shift keying
M-FSK	Multiple frequency-shift keying
MGF	Moment generating function
M-PSK	Multiple phase-shift keying
M-QAM	M-ary quadrature amplitude modulation
MIMO	Multiple input/multiple output
MMSE	Minimum mean-square error
MRC	Maximal-ratio combining
MS	Mobile station
MSK	Minimum-shift keying
$Mx(s)$	Moment generating function of γ
N	Noncoherent
NCFSK	Noncoherent frequency-shift eying

NOTATION	DESCRIPTION OR NAME OF FUNCTION AND REFERENCE CITATION FOR ITS DEFINITION
NT	Normalized threshold
OC	Optimum combining
OPRA	Optimum power and rate adaptation
ORA	Optimum rate adaptation
Pb (E)	Bit error probability
PDF	Probability density function
PG	Processing gain
PLL	Phase-locked loop
Ps (E)	Symbol error probability
QAM	Quadrature amplitude modulation
QASK	Quadrature amplitude-shift keying
QPSK	Quadrature phase-shift keying
RV	Random variable
SC	Selection combining
SEC	Switch-and-examine combining
SEP	Symbol error probability
SER	Symbol error rate
SIMO	Single input/multiple output
SINR	Signal-to-interference plus noise ratio
SIR	Signal-to-interference ratio
SNR	Signal-to-noise ratio
SQAM	Square quadrature amplitude modulation
TIFR	Truncated channel inversion with fixed rate
Var (X)	Variance operator
x	Instantaneous fading SNR
β	Average fading SNR
$\Gamma(x)$	Gamma function [1, Eq. (8.310.1)]
$\gamma(a, x)$	Incomplete gamma function [1, Eq. (8.350.1)]
$\Gamma(a, x)$	Complementary incomplete gamma function [1, Eq. (8.350.2)]
$I_\nu(x)$	νth-order modified Bessel function of the first kind [1, Eq. (8.431)]
$\psi(a, b, x)$	Confluent hypergeometric function of the second kind [1, Eq. (3.383.5^7)]
ρ	Correlation coefficient
Ω	Average value of signal power

Reference

1. Gradshteyn, I. and Ryzhik, I. (1980). Tables of Integrals, Series, and Products. Academic Press, New York.

Author

 **Stefan R. Panić** received his MSc and PhD in electrical engineering from the Faculty of Electronic Engineering, Niš, Serbia, in 2007 and 2010, respectively. His research interests in the field of mobile and multichannel communications include statistical characterization and modeling of fading channels, performance analysis of diversity combining techniques, and outage analysis of multiuser wireless systems subject to interference. In the field of digital communications, his current research interests include information theory, source and channel coding, and signal processing. He has published more than 40 SCI indexed papers. Currently, he works as docent in the Department of Informatics, Faculty of Natural Science and Mathematics, University of Priština, Serbia.

1

INTRODUCTION

The continuous development of various wireless communication system services leads to a permanent necessity of analyzing the possibility of their performance improvement. Unfortunately, signal propagation in the wireless medium is accompanied by various side effects and drawbacks such as multipath fading and shadowing. Mathematical characterization of these complex phenomena, which describes various types of propagation environments, has been presented in Chapter 2. First, various models, already known in the literature, such as Rayleigh, Rician, Hoyt, Nakagami-m, Weibull, α-μ, η-μ, and κ-μ fading models, used for the statistical modeling of multipath influence, have been introduced. Then, some models for statistical modeling of shadowing influence, such as log-normal and gamma shadowing model, are presented. Finally, composite models are discussed, which correspond to the scenario when multipath fading is superimposed on shadowing. Suzuki, Rician-shadowing, and generalized K composite fading models have been discussed. Further analysis has been extended in Chapter 3, where some correlative fading models have been introduced, considering exponential, constant, and general types of correlation between random processes. In Chapter 4, several performance measures related to the wireless communication system design, such as average signal-to-noise ratio, outage probability, average symbol error probability, amount of fading, level crossing rate, and average fade duration, have been defined and mathematically modeled. Basic concepts of several space diversity reception techniques, such as maximal ratio combining, equal gain combining, selection combining, and switch-and-stay combining, have also been portrayed, with emphasis on the evaluation of reception performance measures. Necessity and the validity of space diversity technique usage, from the point of view of multipath fading and CCI influence mitigation, have been shown in Chapter 5, where single-channel receiver performances have been evaluated for a few general propagation models. Performance improvement at the receiver, achieved by application of diversity reception techniques through the standard performance criteria, has been considered in Chapter 6. Cases of uncorrelated multichannel reception, like diversity reception cases over correlated fading channels, have also been analyzed. Necessity and validity of the macrodiversity reception usage, from the point of view of multipath fading and shadowing mitigation through the second-order statistical measures at the output of the macrodiversity receiver, have been considered in Chapter 7. Finally, in Chapter 8, channel capacity analysis under various adaptation policies and diversity techniques over fading channels has been provided.

As already mentioned in the Preface, this work could help extension of subjects that are normally covered in standard Digital Communication and Signal Processing courses. However, since applications chosen for analyzing correspond to practical systems, the performance study provided in this book will have far more than academic value. The presented collection of system performances will help researchers and system designers to perform trade-off studies among the various communication type/drawback combinations in order to determine the optimal choice in the presence of their available constraints.

2

MODELING OF FADING CHANNELS

Wireless communication channel propagation is a complex phenomenon influenced by various drawbacks such as multipath fading and shadowing. The exact mathematical characterization of these phenomena arises in a very complex form, so system analysis is complicated from this point of view. Nevertheless, significant attempts have been made so far to discover simple and accurate statistical models that describe various types of propagation environments.

During wireless transmission, the envelope and phase of the observed signal fluctuate over time. Phase variations, caused by various factors, could be insignificant since for some digital modulations (noncoherent modulations), phase information is not taken into account at the receiver. However, there is a set of digital modulation schemes (coherent) where phase information is needed at the receiver, so unless some techniques are performed at the reception, phase variations could seriously degrade system performances. However, the emphasis in this work will be on envelope statistic consideration.

The first division of fading types is based on how channel impulse response changes within the symbol duration. If the symbol time duration is smaller than the time duration, after which the correlation function of two samples of the channel response taken at the same frequency but different time instants falls under the defined value (coherence time), then the fading is considered to be slow (long-term, shadowing), contrary to a fast fading (short-time) case with a smaller value of coherence time. The second division is on frequency-flat (nonselective) fading channels and frequency-selective fading channels. In narrowband systems, the signal bandwidth is smaller than the coherence bandwidth, and all spectral components are affected by the same effects, so fading is flat, contrary to wideband signals where fading is selective, since the signal bandwidth is bigger than the coherence bandwidth of the channel.

2.1 Multipath Fading

Multipath fading is caused by atmospheric ducting, ionosphere refraction, and reflection from various objects, so randomly delayed, reflected, scattered, and diffracted signal components combine in a constructive or destructive manner. Multipath fading causes short-term signal variations, and its influence on the signal envelope has been statistically modeled by various models so far. Some of them are presented in the next subsections.

2.1.1 Rayleigh Fading Model

Let us consider two zero-mean statistically independent normally (Gaussian) distributed random variables X_1 and X_2, each with a variance σ^2. Then,

$$r = \sqrt{X_1^2 + X_2^2} \tag{2.1}$$

would represent a Rayleigh distributed random variable, with PDF (probability density function) given in the form of [1]

$$f_R(r) = \frac{2r}{\Omega} \exp\left(-\frac{r^2}{\Omega}\right) \tag{2.2}$$

with $\Omega = 2\sigma^2$ representing the average value of the signal power, that is, $\Omega = E(r^2)$, where E denotes the mathematical expectation of the statistical process.

For extensive measurements of the received signal envelope [2,3] in urban and suburban areas, where the line-of-sight (LOS) component is often blocked by various obstacles, the Rayleigh distribution model was proposed as a suitable envelope model process. Rayleigh fading distribution is sometimes called the scatter or diffuse distribution.

If the received carrier amplitude is modulated by the fading amplitude r, where r is a random variable with mean-square value Ω, and probability density function (PDF), $f_R(r)$, which is dependent on the nature of the wireless propagation environment, after passing through the fading channel, the signal is perturbed at the receiver by additive white Gaussian noise (AWGN), which is typically assumed to be statistically independent of the fading amplitude r and is characterized by a one-sided power spectral density N_0 (W/Hz) [1]. Then the instantaneous signal-to-noise power ratio (SNR) per symbol is defined by $x = r^2 Es/N_0$ and the average SNR per symbol by $\beta = \Omega\, Es/N_0$, where E_s is the energy per symbol [1]. Now, from the previous expression it follows that $x = r^2\, \beta/\Omega$, and the PDF of x is obtained by introducing a change of variables in the expression for the fading amplitude PDF yielding

$$f_X(x) = f_R\left(\sqrt{x\frac{\Omega}{\beta}}\right)|J|; \quad |J| = \frac{dr}{dx} = \frac{1}{2\sqrt{x(\beta/\Omega)}}. \tag{2.3}$$

Then, in the case of Rayleigh distributed envelope, SNR PDF is exponentially distributed:

$$f_X(x) = \frac{1}{\beta}\exp\left(-\frac{x}{\beta}\right). \tag{2.4}$$

Based on which performance evaluation of wireless communication parameters could be carried out, the other two important first-order signal statistic measures are the cumulative distribution function (CDF) and the moment generation function (MGF). For Rayleigh distribution cases, they could be associated with the amplitude and SNR PDFs and defined as

$$F_R(r) = \int_0^r f_T(t)dt = 1 - \exp\left(-\frac{r^2}{\Omega}\right); \tag{2.5}$$

and

$$M_X(s) = \int_0^\infty f_X(x)\exp(-xs)dx = \frac{1}{1-s\beta}. \tag{2.6}$$

2.1.2 Rician Fading Model

Let us start from two zero-mean, statistically independent normally (Gaussian) distributed random variables X_1 and X_2, each with a variance σ^2. Then

$$r = \sqrt{\left(X_1 + A\right)^2 + X_2^2} \qquad (2.7)$$

would represent a Rician distributed random variable, with PDF given in the form of [4]

$$f_R(r) = \frac{r}{\sigma^2} \exp\left(-\frac{r^2 + A^2}{2\sigma^2}\right) I_0\left(\frac{rA}{\sigma^2}\right), \qquad (2.8)$$

where

A represents the average value of signal power in LOS components

$2\sigma^2$ represents the average value of signal power in non-LOS components

$I_0(x)$ denotes the zeroth order modified Bessel function of the first kind [5, Eq. (8.445)]

Rician distribution is often described in terms of a fading parameter K, which defines the ratio of signal power in the dominant component of the desired signal and the scattered power:

$$f_R(r) = \frac{2(1+K)r}{e^K \Omega} \exp\left(-\frac{(1+K)r^2}{\Omega}\right)$$

$$\times I_0\left[2\sqrt{\frac{K(1+K)r^2}{\Omega}}\right], \quad K = \frac{A^2}{2\sigma^2} = \frac{A^2}{\Omega}. \qquad (2.9)$$

The Rician fading model can be used as a stochastic model for wireless propagation anomaly caused by partial cancellation of a radio signal by itself: the signal arrives at the receiver by two different paths, while at least one of the paths is changing (lengthening or shortening). The Rician fading model is used for modeling propagation paths, consisting of one strong direct line-of-sight (LoS) signal and many randomly reflected signals that are usually weaker signals. This fading model is applicable for modeling the fading channels in the frequency domain [6].

Particularly, considering mobile satellite communications, the Rician fading is used to accurately model the mobile satellite channel for single [7], clear-state [8] channel conditions.

According to (2.3), the PDF of Rician distributed SNR per symbol of the channel (x) is distributed as

$$f_X(x) = \frac{(1+K)}{\exp(K)\beta} \exp\left(-\frac{(1+K)x}{\beta}\right) I_0\left(2\sqrt{\frac{K(1+K)x}{\beta}}\right). \quad (2.10)$$

It can be shown that the amplitude CDF can be presented in the closed form as

$$F_R(r) = \int_0^r f_T(t)\,dt = \sum_{k=0}^{\infty} \frac{A\exp\left(-A^2/2\sigma^2\right)\gamma\left(k+1, r^2/2\sigma^2\right)}{k!\,\Gamma\left(k+1\right)2^k\sigma^{2k}}, \quad (2.11)$$

where $\Gamma(z)$ and $\gamma(a,z)$ denote gamma and lower incomplete gamma function [5, Eq. (8.310⁷.1)]. Similarly, SNR MGF can be presented in the form of

$$M_X(s) = \int_0^\infty f_X(x)\exp(-xs)\,dx = \frac{(1+K)}{(1+K)-s\beta}\exp\left(\frac{Ks\beta}{(1+K)-s\beta}\right). \quad (2.12)$$

2.1.3 Hoyt Fading Model

Let us consider two zero-mean normally (Gaussian) distributed random in-phase and quadrature signal components X_1 and X_2, with arbitrary variances $\sigma_{X_1^2}$ and $\sigma_{X_2^2}$. Then,

$$R = \sqrt{X_1^2 + X_2^2} \quad (2.13)$$

would represent Hoyt (Nakagami-q) distributed random processes, with PDF given in the form of [9]

$$f_R(r) = \frac{(1+q^2)r}{q\Omega}\exp\left(-\frac{(1+q^2)^2 r^2}{4q^2\Omega}\right) I_0\left(\frac{(1-q^4)r^2}{4q^2\Omega}\right), \quad (2.14)$$

where

$\Omega = E[r^2]$ denotes the desired signal average power

$q = \sigma_X/\sigma_Y$, $0 \le q \le 1$ is the desired signal Hoyt fading parameter

Hoyt distribution spans the range from one-sided Gaussian fading ($q = 0$, worst-case fading) to Rayleigh fading ($q = 1$, case of identical variances). The Hoyt process model is typically observed on satellite links subject to strong ionospheric scintillation [10]. Recently [11], the Hoyt process model has been used along with the Rice process for the two-state mobile satellite propagation channel modeling. Phase crossing statistics of the Hoyt channel have also been observed in [12].

Another representation of the Hoyt fading model in the form of parameter b, also known in the literature, is [13]

$$f_R(r) = \frac{2r}{\Omega\sqrt{1-b^2}}\exp\left(-\frac{r^2}{(1-b^2)\Omega_d}\right)I_0\left(\frac{br^2}{(1-b^2)\Omega_d}\right), \quad b = \frac{1-q^2}{1+q^2}.$$

(2.15)

Capitalizing on (2.3), it can be shown that, for the Hoyt model, the SNR per symbol of the channel (x) has the following distribution:

$$f_X(x) = \frac{(1+q^2)x}{2q\beta}\exp\left(-\frac{(1+q^2)^2 x}{4q^2\beta}\right)I_0\left(\frac{(1-q^4)x}{4q^2\beta}\right).$$

(2.16)

The CDF of the Hoyt random amplitude process can be obtained based on (2.14) in the form of

$$F_R(r) = \sum_{k=0}^{\infty}\frac{(1-q^4)^{2k}q}{(1+q^2)^{4k+1}2^{2k-1}\Gamma(k+1)k!}\gamma\left(2k+1,\frac{(1+q^2)^2 r}{4q^2\Omega}\right).$$

(2.17)

MGF corresponding to (2.16) can be presented in the form of

$$M_X(S) = \frac{1}{\sqrt{1-2s\beta+\left(2s\beta q/(1+q^2)\right)^2}}.$$

(2.18)

2.1.4 Nakagami-m Fading Model

Let X_{1i} and X_{2i}, $i = 1,\ldots, m$ be zero-mean statistically independent Gaussian distributed random in-phase and quadrature signal components. Then, the square root of their sum

$$r = \sqrt{\sum_{i=1}^{m}\left(X_{1i}^2 + X_{2i}^2\right)} \qquad (2.19)$$

would represent Nakagami-m distributed random processes, with PDF given in the form of [14]

$$f_R(r) = \frac{2r^{2m-1}m^m}{\Gamma(m)\Omega^m}\exp\left(-\frac{r^2}{\Omega}\right) \qquad (2.20)$$

with $\Omega = E\ (r^2)$ being the average signal power, and m denoting the inverse normalized variance of r^2, which has to satisfy $m \geq 1/2$, describing the fading severity [15], since channel converges to a static channel when $m \to \infty$ [16].

Since the Nakagami-m fading model describes multipath scattering with large delay-time spreads, and different clusters of reflected waves, it provides good fits to collected data in indoor and outdoor wireless environments [16]. Recent studies have proved also that the Nakagami-m fading model gives the best fit for satellite-to-indoor and satellite-to-outdoor radio wave propagation [17]. Similarly, best fit to land-mobile and indoor-mobile multipath propagation as well as scintillating ionospheric radio links [15] can be obtained by observing the Nakagami-m fading model.

As a general fading distribution, the Nakagami-m fading model includes (as special cases) other distributions such as Rayleigh distribution (by setting parameter m value $m = 1$) and one-sided Gaussian distribution ($m = 1/2$). In addition, for more severe scenarios than Rayleigh fading $m < 1$, the Nakagami-m distributed fading closely approximates the Hoyt distribution, by introducing a relation between parameters q and m, as follows [1]:

$$m = \frac{(1+q^2)^2}{2(1+2q^4)}. \qquad (2.21)$$

Finally, for less severe scenarios $m > 1$, the Nakagami-m distribution could approximate the Rician distribution, by introducing a relation between the mentioned parameters [1]:

$$m = \frac{(1+K)^2}{1+2K}. \qquad (2.22)$$

Capitalizing on (2.3), the PDF of Nakagami-m distributed SNR per symbol of the channel (x) is gamma distributed:

$$f_X(x) = \frac{x^{m-1} m^m}{\Gamma(m) \beta^m} \exp\left(-\frac{x}{\beta}\right) \tag{2.23}$$

with corresponding MGF in the form of

$$M_X(s) = \int_0^\infty f_X(x) \exp(-xs)\, dx = \frac{1}{(1 - s\beta)^m}. \tag{2.24}$$

Starting from (2.20), it can be shown that the CDF of the Nakagami-m random amplitude process can be presented in the form of

$$F_R(r) = \int_0^r f_T(t)\, dt = \gamma\left(m, m\frac{r^2}{\Omega}\right). \tag{2.25}$$

2.1.5 Weibull Fading Model

Let us consider zero-mean normally (Gaussian) distributed random in-phase and quadrature variables X_1 and X_2 with equal variances. The resulting envelope, obtained as a nonlinear function of their sum,

$$r = \sqrt[\alpha]{X_1^2 + X_2^2}, \tag{2.26}$$

follows the Weibull distribution, with the PDF given in the form of [18]

$$f_R(r) = \frac{\alpha r^{\alpha-1}}{\Omega} \exp\left(-\frac{r^\alpha}{\Omega}\right), \tag{2.27}$$

with α denoting the nonlinearity parameter, while $\Omega = E(r^\alpha)$ is the average signal power. Nonlinearity is introduced to model some propagation phenomenon, which are still under scientific observations [19]. The nonlinearity parameter α is also a measure of fading severity ($\alpha \geq 0$), because when its value increases, the severity of the fading decreases [20]. For the special case of $\alpha = 2$, the Weibull distribution reduces to Rayleigh, while for $\alpha = 1$, it models the exponential distribution.

Experimental data supporting the appropriateness of the Weibull model usage have been confirmed by observing experimental results from [21]. It was used as a model for indoor propagation in [22]. In addition, appropriateness for modeling the path-loss model of a narrow-band digital enhanced cordless telecommunications (DECT) system at reference frequency 1.89 GHz, by the Weibull distribution, is shown in [23]. Finally, the Weibull distribution can be used to model outdoor multipath fading by considering fading channel data obtained from a recent measurement program at 900 MHz [24].

The PDF of the Weibull distributed SNR per symbol of the channel (x) can be obtained based on (2.3) in the form of

$$f_X(x) = \frac{\alpha x^{(\alpha/2)-1}}{2\beta} \exp\left(-\frac{x^{(\alpha/2)}}{\beta}\right). \tag{2.28}$$

It can be shown that the CDF of the Weibull random amplitude process can be presented in the form of

$$F_R(r) = \int_0^r f_T(t)\, dt = 1 - \exp\left(-\frac{r^\alpha}{\Omega}\right). \tag{2.29}$$

According to [25], SNR MGFs can be presented in the form of

$$M_X(S) = \int_0^\infty f_X(x)\exp(-xs)\, dx = \frac{\alpha(2\pi)^{(1-\alpha)/2}\alpha^{(1-\alpha)/2}}{\beta s^\alpha}$$

$$\times G_{1,\alpha}^{\alpha,1}\left[\beta\left(\frac{\alpha}{s}\right)^\alpha \middle| \begin{matrix} 1 \\ 1,\ldots,1+(\alpha-1)/\alpha \end{matrix}\right], \tag{2.30}$$

where $G_{p,q}^{m,n}\left[\begin{matrix}(a)_p \\ (b)_q\end{matrix} \middle| x\right]$ is the Meijer's G-function [5, Eq. (9.301[7])].

2.1.6 α-μ (Generalized Gamma) Fading Model

Let us consider the case when the received signal encompasses an arbitrary number of in-phase and in-quadrature multipath components, modeled with zero-mean Gaussian distributed random variables

X_{1i} and X_{2i}, $i = 1, \ldots, \mu$, with equal variances. Then the resulting envelope obtained as a nonlinear function of their sum

$$r = \sqrt[\alpha]{\sum_{i=1}^{\mu} \left(X_{1i}^2 + X_{2i}^2 \right)} \qquad (2.31)$$

follows α-μ distribution, with the PDF given in the form of [26]

$$f_R(r) = \frac{\alpha \mu^{\mu} r^{\alpha\mu-1}}{\hat{r}^{\alpha\mu} \Gamma(\mu)} \exp\left(-\mu \frac{r^{\alpha}}{\hat{r}^{\alpha}} \right), \qquad (2.32a)$$

where

$$\hat{r}^{\alpha} = E(r^{\alpha}); \quad E(X_{1i}^{2}) = E(X_{2i}^{2}) = \hat{r}^{\alpha}/2\mu$$

α denotes the parameter corresponding to the form of nonlinearity

This model provides a very good fit to the measured data over a wide range of fading conditions. This distribution has the same functional form as the generalized gamma or Stacy distribution [26]. Two important phenomena inherent to wireless propagation, namely nonlinearity and clustering, can be explained by the α-μ fading model, since roughly speaking, parameter α models the nonlinearity of the environment, while parameter μ is associated with the number of multipath clusters [27,28]. The α-μ distribution considers a signal composed of clusters of multipath waves propagating in a nonhomogeneous environment. It is assumed that (1) phases of scattered waves within a single cluster are random, but with similar delay times; (2) delay-time spreads of different clusters are relatively large; and (3) clusters of multipath waves have scattered waves with identical powers. As stated in [27], noninteger values of the parameter μ, which have been found in practice, may be on account: (1) nonzero correlation among the clusters of multipath components, (2) nonzero correlation between the in-phase and quadrature components, and (3) non-Gaussianity of the in-phase and quadrature components of the fading signal and others.

The α-μ distribution is a general fading distribution model that includes other famous distributions as special cases, such as the Weibull distribution (can be obtained by setting parameter value μ = 1) and the Nakagami-m distribution (can be obtained by setting parameter value α = 2). Further, one-sided Gaussian and Rayleigh distributions, which are also special cases, could be obtained now from Weibull and Nakagami-m distributions.

It can be shown, from (2.32a), that the CDF of α-μ random amplitude process can be presented in the form of

$$F_R(r) = \int_0^r f_T(t)\,dt = \frac{\gamma\big(\mu, \mu(r^\alpha/\hat{r}^\alpha)\big)}{\Gamma(\mu)}. \qquad (2.32b)$$

Based on (2.3), the PDF of the α-μ distributed SNR per symbol of the channel (x) is distributed as

$$f_X(x) = \frac{\alpha x^{(\alpha\mu/2)-1}}{2\Gamma(\mu)(\Xi\beta)^{\alpha\mu/2}}\exp\left(-\left(\frac{x}{\Xi\beta}\right)^{\alpha/2}\right); \quad \Xi = \frac{\Gamma(\mu)}{\Gamma\big(\mu+(2/\alpha)\big)}. \qquad (2.33)$$

The corresponding MGF is in the form of [29]

$$M_X(s) = \int_0^\infty f_X(x)\exp(-xs)\,dx = \frac{\alpha}{2\Gamma(\mu)(\Xi\beta s)^{\alpha\mu/2}}\frac{k^{1/2}l^{(\alpha\mu-1)/2}}{(2\pi)^{(l+k-2)/2}}$$

$$\times G_{l\,k}^{k\,l}\left(\frac{l^l}{s^l k^k (\Xi\beta s)^{\alpha\mu/2}}\left|\begin{array}{c}\Delta(l, 1-(\alpha\mu/2))\\\Delta(k, 0)\end{array}\right.\right), \qquad (2.34)$$

where $\Delta(k; a)$ is defined as $\Delta(k; a)$, a/k, $(a+1)/k$, ..., $(a+k-1)/k$, with a an arbitrary real value and k as a positive integer. Moreover, k and l are positive integers, so that $l/k = \alpha/2$ holds. Depending on the specific value of α, a set of minimum k and l can be chosen properly.

2.1.7 κ-μ Fading Model

Let X_{1i} and X_{2i} be mutually independent Gaussian processes, with variances $E(X_{1i}) = E(X_{2i}) = \sigma^2$. Modeling mean values of in-phase and quadrature components of the multipath waves of ith cluster, with p_i and q_i, and assuming μ initially to be an integer, the resulting envelope obtained as a function

$$r = \sqrt{\sum_{i=1}^{\mu}\big(X_{1i}+p_i\big)^2 + \sum_{i=1}^{\mu}\big(X_{2i}+q_i\big)^2} \qquad (2.35)$$

follows the κ-μ distribution, with the PDF given in the form of [30]

$$f_R(r) = \frac{2\mu(1+\kappa)^{(\mu+1)/2} r^\mu}{\kappa^{(\mu-1)/2} e^{\mu\kappa}\Omega^{(\mu+1)/2}} \exp\left(-\frac{\mu(1+\kappa)r^2}{\Omega}\right) I_{\mu-1}\left[2\mu\sqrt{\frac{\kappa(1+\kappa)r^2}{\Omega}}\right];$$

$$p^2 = \sum_{i=1}^{\mu} p_i^2; \quad q^2 = \sum_{i=1}^{\mu} q_i^2; \quad \kappa = \frac{p^2+q^2}{2\mu\sigma^2}; \quad \Omega = 2\mu\sigma^2(1+\kappa); \qquad (2.36)$$

where $\Omega = E[R^2]$ is the desired signal average power.

The κ-μ distribution fading model corresponds to a signal composed of clusters of multipath waves. The phases of the scattered waves are random and have similar delay times, within a single cluster, while delay-time spreads of different clusters is relatively large [31]. It is assumed that the clusters of multipath waves have scattered waves with identical powers, and each cluster has a dominant component with arbitrary power. This distribution is well suited for line-of-sight (LoS) applications, since every cluster of multipath waves has a dominant component (with arbitrary power). The κ-μ distribution is a general physical fading model, which includes Rician and Nakagami-*m* fading models as special cases (as the one-sided Gaussian and the Rayleigh distributions since they also represent special cases of Nakagami-*m*). Parameter κ represents the ratio between the total power of dominant components and the total power of scattered components. Parameter μ is related to multipath clustering. As μ decreases, the fading severity increases. For the case of κ = 0, the κ-μ distribution is equivalent to the Nakagami-*m* distribution. When μ = 1, the κ-μ distribution becomes the Rician distribution, with κ as the Rice factor. Moreover, the κ-μ distribution fully describes the characteristics of the fading signal in terms of measurable physical parameters. The fading experienced in body to body 2.45 GHz communication channels for fire and rescue personnel has been characterized in [32] with the κ-μ model, with parameters κ = 2.31 and μ = 1.19, while an excellent fit to measured data is obtained.

Starting from (2.36), the CDF of the κ-μ random amplitude process can be presented in the form of

$$F_r(r) = \int_0^r f_T(t)\,dt = \sum_{p=0}^{+\infty} \frac{\kappa^p \mu^p}{\exp(\kappa\mu)\Gamma(\mu+p)p!} \gamma\left(\mu+p, \frac{\mu(\kappa+1)}{\Omega}r^2\right).$$

(2.37)

The PDF of the κ-μ distributed SNR per symbol of the channel (x) is distributed as

$$f_X(x) = \frac{\mu(1+\kappa)^{(\mu+1)/2}}{\kappa^{(\mu-1)/2}e^{\mu\kappa}\beta^{(\mu+1)/2}} x^{(\mu-1)/2} \exp\left(-\frac{\mu(1+\kappa)x}{\beta}\right)$$

$$\times I_{\mu-1}\left[2\mu\sqrt{\frac{\kappa(1+\kappa)x}{\beta}}\right].$$

(2.38)

In [33], it has been shown that the corresponding MGF can be presented as

$$M_x(s) = \frac{\exp\left(-s(p^2+q^2)/(1+2s\sigma^2)\right)}{(1+2s\sigma^2)^\mu}.$$

(2.39)

2.1.8 η-μ Fading Model

Let X_{1i} and X_{2i}, $i = 1, \ldots, \mu$ be Gaussian in-phase and quadrature zero mean processes, with $E(X_{1i}) = E(X_{2i}) = 0$. Assuming μ initially to be an integer, the resulting envelope obtained as a function

$$r = \sqrt{\sum_{i=1}^{\mu}\left(X_{1i}^2 + X_{2i}^2\right)}$$

(2.40)

follows the η-μ distribution, with PDF given in the form of [34]

$$f_R(r) = \frac{4\sqrt{\pi}\mu^{\mu+1/2}h^\mu r^{2\mu}}{\Gamma(\mu)H^{\mu-1/2}\Omega^{\mu+1/2}} e^{-2\mu hr^2/\Omega} I_{\mu-1/2}\left[\frac{2\mu Hr^2}{\Omega}\right],$$

(2.41)

with $\Omega = E(R^2)$ being the average signal power, while H and h are distribution parameters, which can be defined according to the observed format, since the η-μ distribution covers two formats:

Format 1 and Format 2. In Format 1, the variances of independent Gaussian in-phase and quadrature processes are arbitrary with their ratio defined as $\eta = E(X_{1i}^2)/E(X_{2i}^2)$. Then, distribution parameters are defined in the following manner:

$$H=\frac{\eta^{-1}-\eta}{4}; \quad b=\frac{2+\eta^{-1}+\eta}{4}; \quad \eta \geq 0. \tag{2.42}$$

In Format 2, the variances of dependent Gaussian in-phase and quadrature processes are identical with the correlation coefficient defined as $\eta = E(X_{1i} X_{2i})/E(X_{2i}^2)$ or $\eta = E(X_{1i} X_{2i})/E(X_{1i}^2)$. Then, distribution parameters are defined in the following manner:

$$H=\frac{\eta}{1-\eta^2}; \quad b=\frac{1}{1-\eta^2}; \quad -1 \leq \eta \leq 1. \tag{2.43}$$

Transformation from Format 1 into Format 2 can be obtained by using relation [34]:

$$\eta_{Format1}=\frac{1-\eta_{Format2}}{1+\eta_{Format2}}. \tag{2.44}$$

In wireless communications, the η-μ fading distribution has been used, as a general fading distribution, for representing small-scale signal variation in general non-line-of-sight conditions. Roughly, the parameter μ is related to the number of multipath clusters in the environment.

As a general distribution, this model includes some classical fading distributions as particular cases, e.g., Nakagami-q (Hoyt), one-sided Gaussian, Rayleigh, and Nakagami-m [35]. The Hoyt (or Nakagami-q) distribution can be obtained from the η-μ distribution, in an exact manner, by setting $\mu = 0.5$. In this case, the Hoyt (or Nakagami-q) parameter is given by

$$b=\frac{\eta_{Format1}-1}{\eta_{Format1}+1}; \quad q^2=\eta_{Format1};$$

$$b=-\eta_{Format2}; \quad q^2=\frac{1-\eta_{Format2}}{\eta_{Format2}+1}. \tag{2.45}$$

The Nakagami-m distribution can be obtained from the η-μ distribution, in an exact manner, by setting $\mu = m$ and $\eta \to 0$ (or $\eta \to \infty$) in

Format 1, or $\eta \to \pm 1$ in Format 2. In addition, it can be obtained [36] by setting $\mu = m/2$ and $\eta \to 1$ in Format 1 or $\eta \to 0$ in Format 2.

After integrating, the CDF of the η-μ random amplitude process can be presented in the form of

$$F_R(r) = \int_0^r f_T(t)\,dt = \sum_{p=0}^{+\infty} \frac{\sqrt{\pi}H^{2p}\gamma\left(4p+4\mu, 2\mu h r^2/\Omega\right)}{2^{2p+2\mu-1}\Gamma(\mu)\Gamma(\mu+p+1/2)\,p!\,h^{\mu+2p}}. \quad (2.46)$$

After performing the transformation explained in Section 2.1, the PDF of the η-μ distributed SNR per symbol of the channel (x) can be presented in the following form:

$$f_X(x) = \frac{2\sqrt{\pi}\mu^{\mu+1/2}h^{\mu}x^{\mu-1/2}}{\Gamma(\mu)H^{\mu-1/2}\beta^{\mu+1/2}}\exp\left(-\frac{2\mu h x}{\beta}\right)I_{\mu-1/2}\left[\frac{2\mu H x}{\beta}\right]. \quad (2.47)$$

Based on (2.47), it has been shown in [37] that the MGF can be presented in the following form:

$$M_x(s) = \frac{1}{\left(1+\dfrac{\beta s}{2\mu(h-H)}\right)^{\mu}\left(1+\dfrac{\beta s}{2\mu(h+H)}\right)^{\mu}}. \quad (2.48)$$

2.2 Shadowing (Long Time Fading)

2.2.1 Log-Normal Shadowing Model

According to empirical measurements, a general consensus arises that shadowing should be most accurately modeled by a log-normal (LN) distribution model, in various outdoor and indoor cases. It has been said in [38] that a lot of experimental measurements have shown that the LN distribution is the sovereign model for describing the shadowing phenomenon, due to large obstructions on outdoor, in-door, and satellite channels. It is also well known [1] that the small- and large-scale effects get mixed in slowly varying channels, so that the LN statistics accurately describes the distribution of the channel path gain.

However, the LN model could be analytically very intractable, especially its MGF expression. It has been shown in [1] that the PDF of the SNR ratio per symbol x could be expressed by

$$f_x(x) = \frac{10}{\ln 10 \sqrt{2\pi}\sigma x} \exp\left(-\frac{(10 \log x - \mu)^2}{2\sigma^2}\right), \tag{2.49}$$

with μ(dB) and σ(dB) being the mean and the standard deviation of $10 \log x$, respectively (shadowing spread, typical values in macrocells range from 5 to 12 dB and from 4 to 13 dB in microcells). μ can be estimated by using [39] as

$$\mu = \frac{P_t C}{d^a (1 + d/g)^b}, \tag{2.50}$$

where
   P_t stands for the transmitted power
   C describes the effects of antenna gain
   d denotes the distance between the transmitter and receiver
   g is the break point
   a is the basic path-loss exponent
   b is the additional path-loss exponent

2.2.2 Gamma Shadowing Model

It has already been shown in the literature [40–43] that the gamma distribution, which is analytically better tractable, closely approximates the LN distribution in a wide range of propagation conditions and has a good fit to experimental data.

The PDF of the gamma process SNR ratio per symbol, x, is described by [43]

$$f_X(x) = \frac{x^{c-1}}{\beta^c \Gamma(c)} \exp\left(-\frac{x}{\beta}\right) \tag{2.51}$$

with parameter c denoting the measure of shadowing severity present in the channel. The shadowing level described by parameter c and the shadowing spread σ (dB) of the LN model are related through [44]

$$\sigma(dB) = 4.324 \sqrt{\psi'(c)} \tag{2.52}$$

with $\psi'(.)$ being the Trigamma function [5].

Starting from (2.51), it can be shown that the corresponding MGF could be presented in the form of

$$M_x(s) = \int_0^\infty f_X(x)\exp(-xs)\,dx = \frac{1}{(1-s\beta)^c}. \qquad (2.53)$$

2.3 Composite Fading Models

When multipath fading is superimposed on shadowing, the instantaneous composite multipath/shadowed signal should be analyzed at the receiver. This scenario often occurs in land–mobile satellite systems subjected to vegetative and/or urban shadowing. A few models have been proposed in the literature for describing this phenomenon. Some of them are presented in the next subsection.

2.3.1 Suzuki Fading Model

In order to characterize the urban radio propagation medium in various urban environments, a new Suzuki composite distribution was introduced in [45]. This composite distribution describes the scenario in which the main wave is subjected to a local scattering by the cluster of buildings but also traverses a path subjected to the influence of multiple reflections and/or diffractions by natural and man-made objects. The signal is composed from local subpath signals at the receiver, as a result of scattering from local objects. All subpath signals have roughly the same delay but different carrier phases. The signal is assumed to have a lognormal strength, since it undergoes multiplication effects, while the signal distribution is Rayleigh due to an additive scattering effect.

The PDF of the composite Suzuki process envelope is described by [45]

$$f_X(x) = \int_0^\infty \frac{x}{\beta^2}\exp\left(-\frac{x^2}{2\beta^2}\right)\frac{10}{\ln 10\sqrt{2\pi}\,\beta\lambda}\exp\left(-\frac{(10\log\beta-\mu)^2}{2\lambda^2}\right)d\beta.$$

$$(2.54)$$

The Suzuki process is obtained by multiplication of the Rayleigh process by a log-normal process, and it has been proven that two distributions have the same functional form. The general lack of this distribution appliance is caused by its complicated integral form.

2.3.2 Generalized K Fading Model

As already discussed, composite propagation environments of multipath fading superimposed by lognormal or gamma shadowing, resulting in lognormal or gamma-based fading models, such as Suzuki, Rayleigh-longnormal, Rician-longnormal, or Nakagami-lognormal, and Nakagami-gamma fading channels, are analytically very difficult to handle. Generalized K distribution [46–48] is a general and versatile distribution, which could be used for the accurate modeling of a great variety of composite channel conditions. It has also been shown [49] that the K model provides a similar performance to the well-known Rayleigh-lognormal (Suzuki) model while also being mathematically tractable.

In order to superimpose the influence of multipath fading, modeled by the Nakagami-m distribution with a shadowing process modeled with gamma distribution by using the total probability theorem, after some mathematical manipulations, the PDF of the novel composite distribution can be expressed in closed form, by averaging the short-time conditional RV process as [46–48]

$$f_R(r) = \int\limits_0^\infty f_{R|Y}(r|y) f_Y(y) \, dy;$$

$$f_{R|Y}(r|y) = \frac{2r^{2m-1}m^m}{\Gamma(m)y^m} \exp\left(-\frac{r^2}{y}\right); \quad f_Y(y) = \frac{y^{k-1}}{\Omega^k \Gamma(k)} \exp\left(-\frac{y}{k}\right);$$

$$f_R(r) = \frac{4}{\Gamma(m)\Gamma(k)}\left(\frac{m}{\Omega}\right)^{(m+k)/2} r^{m+k-1} K_{k-m}\left(2r\sqrt{\frac{m}{\Omega}}\right), \tag{2.55}$$

with $K_v(x)$ denoting the modified Bessel function of second kind and order v [5, Eq. (8.407/1)], and $\Omega = E(x^2)/k$. The parameter m is a fading severity parameter and k is a shadowing severity parameter. The two shaping parameters, m and k, can take different values

($m \geq 1/2$ and $k \in (0, \infty)$), which approximate several shadowing conditions, from severe shadowing ($k \to 0$) to no shadowing ($k \to \infty$). Therewith, a great variety of short-term and long-term fading (shadowing) conditions can be described.

After integrating, the CDF of the generalized K random amplitude process can be presented in the form of

$$F_R(r) = \int_0^r f_T(t)\,dt = \frac{r^{(m_d + k_d + 1)/2}}{\Gamma(m_d)\Gamma(k_d)}\left(\frac{m_d}{\Omega_d}\right)^{(m_d + k_d)/2}$$

$$\times G_{1,3}^{2,1}\left(\begin{array}{c} \dfrac{1 - m_d - k_d}{2} \\[2mm] \dfrac{m_d - k_d}{2}, -\dfrac{1 + m_d + k_d}{2}, \dfrac{k_d - m_d}{2} \end{array}\Bigg|\, r\frac{m_d}{\Omega_d}\right).$$

$$(2.56)$$

The PDF of the generalized K distributed SNR per symbol of the channel (x) can be presented in the following form by using the approach presented in [50] as

$$f_X(x) = \frac{\pi c \sec(\pi(m - k))}{\Gamma(m)\Gamma(k)\beta}\left[\left(\frac{mk}{\beta}\right)^m \frac{x^m\,_0F_1(1 + m - k; mkx/\beta)}{\Gamma(1 + m - k)}\right.$$

$$\left. -\left(\frac{mk}{\beta}\right)^k \frac{x^k\,_0F_1(1 - m + k; mkx/\beta)}{\Gamma(1 - m + k)}\right].$$

$$(2.57)$$

Similarly, the MGF can be presented in the following form [50]:

$$M_X(s) = \frac{\pi c \sec(\pi(m - k))}{\Gamma(m)\Gamma(k)\beta}\left[\left(\frac{mk}{\beta}\right)^m \frac{_1F_1(m; 1 + m - k; mk/s\beta)}{s^m\Gamma(1 + m - k)}\right.$$

$$\left. -\left(\frac{mk}{\beta}\right)^k \frac{_1F_1(k; 1 - m + k; mk/s\beta)}{s^k\Gamma(1 - m + k)}\right].$$

$$(2.58)$$

2.3.3 Rician Shadowing Model

This type of channel model is often used in describing performances of terrestrial systems with line-of-sight (LOS) paths and land mobile

satellite (LMS) systems. The random fluctuations of the signal envelope in narrow-band LMS channels can be described depending on the
channel conditions. Multipath fading, caused by the weak scatter components propagated via different non-LOS paths together with the non-
blocked LOS component, is analyzed in [51]. The LOS component is
observed as a Nakagami-m variable. In that way, the power of the LOS
component is a gamma random variable, which is an alternative to the
lognormal distribution and can result in a simpler statistical model with
the same performance for practical cases of interest. In addition, the Rice
model with the Nakagami-distributed LOS amplitude [51] constitutes
a versatile model that not only agrees very well with measured LMS
channel data but also offers significant analytical and numerical advan-
tages for system performance predictions, design issues, etc. The PDF of
the previously described random envelope process is given by [51]

$$f_R(r) = \left(\frac{2bm}{2bm+\Omega}\right)^m \frac{r}{b} \exp\left(-\frac{r^2}{2b}\right) {}_1F_1\left(m,1,\frac{\Omega r^2}{2b(2bm+\Omega)}\right), \quad (2.59)$$

where
  $2b$ is the average power of the scatter component
  Ω is the average power of the LOS component
  m is the Nakagami fading parameter
  ${}_1F_1(a,b,x)$ is the confluent hypergeometric function [5, Eq. (9.210/1)]

In the traditional Nakagami model for multipath fading, m changes
over the limited range of $m \geq 0.5$, while this model allows m to vary
over the wider range of $m \geq 0$ [51]. This enables the evaluated PDF
(2.59) to approximate accurately a great variety of LOS conditions in
LMS channels. For $0 < m < \infty$, suburban and rural areas with partial
obstruction of the LOS can be described. The extreme cases of $m = 0$
or $m = \infty$ refer to the model of urban areas with complete obstruction or open areas with no obstruction of the LOS, respectively. Of
course, these extremes use the infinite series representation of confluent hypergeometric function.

The CDF of the envelope process can be derived as

$$F_R(r) = \int_0^r f_T(t)\,dt = \sum_{k=0}^{\infty} \frac{2^{m-1}b^{m-1}m^m\Omega^k}{(2bm+\Omega)^{m+k}} \frac{\Gamma(m+k)}{(k!)^2\Gamma(m)}\gamma\left(k,\frac{r^2}{2b}\right). \quad (2.60)$$

After performing the transformation explained in Section 2.1, the PDF of the Rician shadowed distributed SNR per symbol of the channel (x) can be presented in the following form:

$$f_X(x) = \left(\frac{2bm}{2bm+\beta}\right)^m \frac{1}{2b} \exp\left(-\frac{x}{2b}\right) {}_1F_1\left(m,1,\frac{\beta x}{2b(2bm+\beta)}\right). \quad (2.61)$$

Finally, the corresponding MGF can be presented as

$$M_X(s) = \frac{(2bm)^m(1-2bs)^{m-1}}{\left[(2bm+\beta)(1-2bs)-\beta\right]^m}. \quad (2.62)$$

References

1. Simon, M. K. and Alouini, M. S. (2005). *Digital Communications over Fading Channels*, 2nd edn. Wiley, New York.
2. Nylund, H. W. (1968). Characteristics of small-area signal fading on mobile circuits in the 150 MHz band. *IEEE Transactions on Vehicular Technology*, 17(1), 24–30.
3. Okumura, Y., Ohmori, E., Kawano, T., and Fukuda, K. (1968). Field strength and its variability in VHF and UHF land mobile radio services. *Review of the Electrical Communication Laboratory*, 16(9–10), 825–873.
4. Rice, S. O. (1948). Statistical properties of a sine wave plus random noise. *Bell System Technical Journal*, 27(1), 109–157.
5. Gradshteyn, I. and Ryzhik, I. (1980). *Tables of Integrals, Series, and Products*. Academic Press, New York.
6. Witrisal, K., Kim, Y. H., and Prasad, R. (2001). A new method to measure parameters of frequency-selective radio channels using power measurements. *IEEE Transactions on Communications*, 49(10), 1788–1800.
7. Corazza, G. E. and Vatalaro, F. (1994). A statistical model for land mobile satellite channels and its application to no geostationary orbit systems. *IEEE Transactions on Vehicular Technology*, 43(3, part 2), 738–742.
8. Wakana, H. (1991). A propagation model for land mobile satellite communications. *Proceedings of the IEEE Antennas and Propagation Society International Symposium*, Amsterdam, the Netherlands, Vol. 3, pp. 1526–1529.
9. Hoyt, R. S. (1947). Probability functions for the modulus and angle of the normal complex variate. *Bell System Technical Journal*, 26(4), 318–359.
10. Chytil, B. (1967). The distribution of amplitude scintillation and the conversion of scintillation. *Journal of Atmospheric and Terrestrial Physics*, 29(9), 1175–1177.
11. Mehrnia, A. and Hashemi, H. (1999). Mobile satellite propagation channel. Part II—A new model and its performance. *Proceedings of the IEEE Vehicular Technology Conference*, Amsterdam, the Netherlands, pp. 2780–2784.

12. Youssef, N., ElBahri, W., Patzold, M., and ElAsmi, S. (2005). On the crossing statistics of phase processes and random FM noise in Nakagami-q mobile fading channels. *IEEE Transactions on Wireless Communications*, 4(1), 24–29.

13. Annamalai, A., Tellambura, C., and Bhargava, V. (2000). Equal-gain diversity receiver performance in wireless channels. *IEEE Transactions on Communications*, 48(10), 1732–1745.

14. Nakagami M. (1964). The *m*-distribution—A general formula of intensity distribution of rapid fading. In *Statistical Methods in Radio Wave Propagation* (W. G. Hoffman ed.), Pergamon Press, Oxford, U.K., pp. 3–36.

15. Simon, M. K., Omura, J. K., Scholtz, R. A., and Levitt. B. K. (1994). *Spread Spectrum Communication Handbook*, revised edition. McGraw-Hill Inc, New York.

16. Rubio, L., Reig, J., and Cardona, N. (2007). Evaluation of Nakagami fading behaviour based on measurements in urban scenarios. *International Journal of Electronics and Communications (AEUE)*, 61(2), 135–138.

17. Lakhzouri, A., Lohan, E., Saastamoinen, I., and Renfors, M. (2005). Measurement and characterization of satellite-to-indoor radio wave propagation channel. *Proceedings of the European Navigation Conference*, Munich, Germany, pp. 1–6.

18. Papoulis, P. (2002). *Probability, Random Variables, and Stochastic Processes*, 4th edn. The McGraw-Hill Companies, New York.

19. Carozzi, T. D. (2002). Radio waves in ionosphere: Propagation, generation, and detection. PhD dissertation, Uppsala University, Sweden.

20. Tzeremes, G. and Christodoulou, C. G. (2002). Use of Weibull distribution for describing outdoor multipath fading. *Proceedings of the IEEE Antennas and Propagation Society International Symposium*, San Antonio, Texas, Vol. 1, pp. 232–235.

21. Shepherd, N. H. (1977). Radio wave loss deviation and shadow loss at 900 MHz. *IEEE Transactions on Vehicular Technology*, 26, 309–313.

22. Hashemi, H. (1993). The indoor radio propagation channel. *Proceedings of the IEEE*, 81, 943–968.

23. Babich, F. and Lombardi, G. (2000). Statistical analysis and characterization of the indoor propagation channel. *IEEE Transactions on Communications*, 48(3), 455–464.

24. Sagias, N. C. and Karagiannidis G. K. (2005). Gaussian class multivariate Weibull distributions: Theory and applications in fading channels. *IEEE Transactions on Information Theory*, 51(10), 3608–3619.

25. Cheng, J., Tellambura, C., and Beaulieu, N. C. (2003). Performance analysis of digital modulations on Weibull fading channels. *Proceedings of the IEEE Vehicular Technology Conference*, Orlando, FL, pp. 236–240.

26. Stacy, E. W. (1962). A generalization of the Gamma distribution. *Annals Mathematical Statistics*, 33(3), 1187–1192.

27. Yacoub, M. D. (2007). The α-μ distribution: A physical fading model for the Stacy distribution. *IEEE Transactions on Vehicular Technology*, 56(1), 27–34.

28. Yacoub, M. D. (2002). The α-μ distribution: A general fading distribution. *Proceedings of the 13th International Symposium on Personal, Indoor, and Mobile Radio Communications*, Lisboa, Portugal, Vol. 2, pp. 629–633.

29. Sagias, N. C. and Mathiopoulos, P. T. (2005). Switched diversity receivers over generalized Gamma fading channels. *IEEE Communications Letters*, 9(10), 871–873.

30. Yacoub, M. D. (2007). The κ-μ distribution and the η-μ distribution. *IEEE Antennas and Propagation Magazine*, 49(1), 68–81.

31. Filho, J. C. and Yacoub, M. D. (2005). Highly accurate κ-μ approximation to sum of M independent non-identical Ricean variates. *Electronics Letters*, 41(6), 338–339.

32. Cotton, S., Scanlon, W., and Guy, J. (2008). The κ-μ distribution applied to the analysis of fading in body to body communication channels for fire and rescue personnel. *IEEE Antennas and Wireless Propagation Letters*, 7, 66–69.

33. Milisic, M., Hamza, M., and Hadzialic, M. (2009). BEP/SEP and outage performance analysis of L-branch maximal-ratio combiner for κ-μ fading. *International Journal of Digital Multimedia Broadcasting*, 2009, Article ID 573404, 1–8.

34. Da Costa, D. B. and Yacoub, M. D. (2007). The η-μ joint phase-envelope distribution. *IEEE Antennas and Wireless Propagation Letters*, 6(1), 195–198.

35. Filho, J. C. and Yacoub, M. D. (2005). Highly accurate η-μ approximation to sum of M independent non-identical Hoyt variates. *IEEE Antennas and Wireless Propagation Letters*, 4(1), 436–438.

36. Da Costa, D. B., Filho J. C., Yacoub, M. D., and Fraidenraich, G. (2008). Second-order statistics of η-μ fading channels: Theory and applications. *IEEE Transactions on Wireless Communications*, 7(3), 819–824.

37. Peppas, K. et al (2010). Sum of non-identical squared η-μ variates and applications in the performance analysis of DS-CDMA systems. *IEEE Transactions on Wireless Communications*, 9(9), 2718–2723.

38. Stuber, G. L. (2001). *Principles of Mobile Communication*, 2nd edn. Kluwer Academic Publishers, Massachusetts.

39. Wang, L. C., Stuber, G. L., and Lea, C. T. (1999). Effects of Rician fading and branch correlation on a local-mean-based macrodiversity cellular system, *IEEE Transactions on Vehicular Technology*, 48(2), 429–436.

40. Abdi, A. and Kaveh, M. (1998). κ-distribution: An appropriate substitute for Rayleigh-lognormal distribution in fading-shadowing wireless channels. *Electronic Letters*, 34(9), 851–852.

41. Abdi, A. and Kaveh, M. (1999). On the utility of Gamma PDF in modeling shadow fading (slow fading). *Proceedings of the IEEE Vehicular Technology Conference*, Amsterdam, the Netherlands, Vol. 3, pp. 2308–2312.

42. Salo, J., Vuokko, L., El-Sallabi, H., and Vainikainen, P. (2007). An additive model as a physical basis for shadow fading. *IEEE Transactions on Vehicular Technology*, 56(1), 13–26.

43. Kostic, P. M. (2005). Analytical approach to performance analysis for channel subject to shadowing and fading. *IEE Proceedings Communications*, 152(6), 821–827.

44. Shankar, P. M. (2007). Outage analysis in wireless channels with multiple interferers subject to shadowing and fading using a compound pdf model. *International Journal of Electronics and Communications (AEUE)*, 61(4), 255–261.
45. Suzuki, H. (1977). A statistical model for urban radio propagation. *IEEE Transactions on Communications*, 25(7), 673–680.
46. Shankar, P. M. (2004). Error rates in generalized shadowed fading channels. *Wireless Personal Communications*, 28(3), 1–6.
47. Iskander, D. and Zoubir, A. (1999). Estimation of the parameters of the K-distribution using higher order and functional moments. *IEEE Transactions on Aerospace and Electronic Systems*, 35(4), 1453–1457.
48. Chitroub, S., Houacine, A., and Sansal, B. (2002). Statistical characterization and modeling of SAR images. *Signal Processing*, 82(1), 69–92.
49. Abdi, A. and Kaveh, M. (2000). Comparison of DPSK and MSK bit error rates for K and Rayleigh-Lognormal fading distributions. *IEEE Communication Letters*, 4(4), 122–124.
50. Bithas, P. S., Sagias, N. C., Mathiopoulos, P.T., Karagiannidis, G. K., and Rontogiannis, A. A. (2006). On the performance analysis of digital communications over generalized-K fading channels. *IEEE Communications Letters*, 10(5), 353–355.
51. Abdi, A., Lau, W. C., Alouini, M. S., and Kaveh, M. (2003). A new simple model for land mobile satellite channels: first- and second-order statistics. *IEEE Transactions on Wireless Communications*, 2(2), 519–528.

3

CORRELATIVE
FADING MODELS

The usage of multichannel receivers is a common feature in various practical realizations of wireless communication systems, caused by a need to combat the effects of various transmission drawbacks [1]. With multichannel reception, it is usually assumed that the received signals are mutually independent. Nevertheless, in a number of practical realizations, this assumption cannot be taken for granted, due to insufficient antenna spacing at small-size mobile terminals (designed according to economical limitations and manufacturing restrictions). A well-known fact is that the correlation in fading across multiple diversity channels results in a degradation of the obtained diversity gain [2–4]. Generally, antenna separation in a distance of a few dozen wavelengths of carrier frequency is typically required to obtain low correlation between branches. Several correlation models have been proposed in the literature so far [2,3,5], and standard performance measures at the reception, considering various detection, modulation, and diversity combining scenarios, have been discussed.

3.1 Novel Representations of Multivariate
Correlative α-μ Fading Model

Considering the α-μ fading model, joint multivariate probability density functions will be derived for various correlation models in this section. These expressions will be further used for the mathematical characterization of multibranch reception analysis over the correlative α-μ fading model. The derivation of expressions is based on using joint multivariate probability density functions obtained for

Nakagami-*m* model. The following relation between the Nakagami-*m* and α-μ random variables (RVs) must be taken into account here [6]:

$$R_{\alpha-\mu}^{\alpha} = R_{Nakagami-m}^{2} \tag{3.1}$$

Let also

- $R_{N_1}, ..., R_{N_n}$ be *n* Nakagami-*m* random variables, with its statistical characterization described with parameters m_1, $E(R_{N_1}^2) = \Omega_1$, ..., m_n, $E(R_{N_n}^2) = \Omega_n$
- R_1, ..., R_n be *n* α-μ random variables, with their statistical characterization described with parameters α_1, μ_1, $\sqrt[\alpha]{E(R_1^\alpha)} = \hat{R}_1$, ..., α_n, μ_n, $\hat{R}_n$
- $0 \le \rho_N \le 1$ stand for the correlation coefficient between Nakagami-*m* random variables, defined with

$$\rho_N = \frac{\text{cov}(R_{N_i}^2, R_{N_j}^2)}{\sqrt{\text{var}(R_{N_i}^2)\,\text{var}(R_{N_j}^2)}}$$

- $0 \le \rho_{\alpha-\mu} \le 1$ stand for the correlation coefficient between α-μ random variables, defined with

$$\rho_{\alpha-\mu} = \frac{\text{cov}(R_i^{\alpha_i}, R_j^{\alpha_j})}{\sqrt{\text{var}(R_i^{\alpha_i})\,\text{var}(R_j^{\alpha_j})}}$$

The expressions for multivariate joint probability density functions of exponentially and constant-correlated α-μ random variables will be derived in the following sections.

3.1.1 Exponential Correlation Model

Let us consider multivariate correlative distribution of *n* random variables. The correlation model is considered to be exponential if elements of the covariance matrix (correlation coefficients between random variables from multivariate distribution) are formed according to [3]

$$\Sigma_{ij} = \begin{cases} 1, & i = j \\ \rho^{|i-j|}, & i \ne j \end{cases} \quad 1 \le i, j \le n. \tag{3.2}$$

The expression for joint probability density function (JPDF), $f_{R_{N_1},\ldots,R_{N_n}}(R_{N_1},\ldots,R_{N_n})$, of n exponentially correlated Nakagami-m random variables, $R_{N_1},\ldots,R_{N_n}$, can be found in the form of [7]

$$f_{R_{N_1},\ldots,R_{N_n}}(R_{N_1},\ldots,R_{N_n})$$

$$= \frac{R_{N_1}^{\ m-1} R_{N_n}^{\ m}}{2^{m-1}\Gamma(m)(1-\rho_N^{\ 2})^{m(n-1)}} \exp\left(-\frac{R_{N_1}^2 + R_{N_n}^2}{2(1-\rho_N^{\ 2})} - g_1 \right)$$

$$\times \prod_{k=1}^{n-1} R_{N_k} \left(\frac{\rho_N}{1-\rho_N^{\ 2}}\right)^{-(m-1)} I_{m-1}\left(\left(\frac{\rho_N}{1-\rho_N^{\ 2}}\right) R_{N_k} R_{N_{k+1}}\right), \qquad (3.3)$$

with $I_n(x)$ being the nth-order-modified Bessel function of the first kind [8, Eq. (8.406)], while

$$g_1 = \begin{cases} 0, & n=2; \\ \left(\rho_N^{\ 2}+1\right)\Big/ 2\left(1-\rho_N^{\ 2}\right) \displaystyle\sum_{k=1}^{n-1} R_{N_k}^2, & n>2. \end{cases} \qquad (3.4)$$

The statements exposed in Section 3.1 imply

$$R_{N_1}^2 = R_1^{\alpha_1},\ldots, R_{N_n}^2 = R_n^{\alpha_n}; \quad m=\mu; \quad \rho_N = \rho_{\alpha-\mu}. \qquad (3.5)$$

Taking into account the previous relations, the JPDF of exponentially correlated α-μ, $f_{R_1},\ldots, _{R_n}(R_1, \ldots, R_n)$, can be presented in the following form [9]:

$$f_{R_1,\ldots,R_n}(R_1,\ldots,R_n) = |J| f_{R_{N_1},\ldots,R_{N_n}}(R_{N_1},\ldots,R_{N_n}), \qquad (3.6)$$

with J as the Jacobian of the transformation given by

$$|J| = \begin{vmatrix} \dfrac{\partial R_{N_1}}{\partial R_1} & \dfrac{\partial R_{N_1}}{\partial R_2} & \cdots & \dfrac{\partial R_{N_1}}{\partial R_n} \\[2ex] \dfrac{\partial R_{N_2}}{\partial R_1} & \dfrac{\partial R_{N_2}}{\partial R_2} & \cdots & \dfrac{\partial R_{N_2}}{\partial R_n} \\[1ex] \vdots & \vdots & \vdots & \vdots \\[1ex] \dfrac{\partial R_{N_n}}{\partial R_1} & \dfrac{\partial R_{N_n}}{\partial R_2} & \cdots & \dfrac{\partial R_{N_n}}{\partial R_n} \end{vmatrix}$$

$$= \frac{\alpha_1\alpha_2\ldots\alpha_n}{2^n} R_1^{(\alpha_1/2)-1} R_2^{(\alpha_1/2)-1}\ldots R_n^{(\alpha_1/2)-1}. \qquad (3.7)$$

After substituting (3.4), (3.5), and (3.7) into (3.6) by using the well-known transformation of the modified Bessel function of the first kind [8, Eq. (8.406)],

$$I_n(x)=\sum_{k=0}^{\infty}\frac{x^{2k+n}}{2^{2k+n}\Gamma(k+n+1)k!},\qquad(3.8)$$

where $\Gamma(a)$ denotes the gamma function [5, Eq. (8.310[7].1)], and the expression for the JPDF of exponentially correlated α-μ random variables is obtained in the form of

$$f_{R_1,\ldots,R_n}(R_1,\ldots,R_n)$$

$$=\sum_{k_1,k_2,\ldots,k_{n-1}=0}^{\infty}\frac{\alpha_1\alpha_2\ldots\alpha_n\rho_{\alpha-\mu}^{2(k_1+k_2+\cdots+k_{n-1})}}{2^{n\mu+2(k_1+k_2+\ldots+k_{n-1})}\Gamma(\mu)\Gamma(\mu+k_1)\ldots\Gamma(\mu+k_{n-1})k_1!\ldots k_{n-1}!}$$

$$\times\frac{R_1^{\alpha_1(\mu+k_1)-1}R_n^{\alpha_n(\mu+k_{n-1})-1}}{\left(1-\rho_{\alpha-\mu}^2\right)^{(n-1)\mu+2(k_1+k_2+\cdots+k_{n-1})}}\exp\left(-\frac{R_1^{\alpha_1}+R_n^{\alpha_n}}{2(1-\rho_{\alpha-\mu}^2)}-G_1\right)\times G_2,$$

$$(3.9)$$

while

$$G_1=\begin{cases}0, & n=2;\\\dfrac{\left(\rho_{\alpha-\mu}^2+1\right)}{2\left(1-\rho_{\alpha-\mu}^2\right)}\displaystyle\sum_{i=2}^{n-1}R_i^{\alpha_i}, & n>2;\end{cases}\qquad G_2=\begin{cases}1, & n=2;\\\displaystyle\prod_{i=2}^{n-1}R_i^{\alpha_i(\mu+k_{i-1}+k_i)-1}, & n>2.\end{cases}$$

$$(3.10)$$

The derived expression could be used for the performance analysis of multibranch space diversity reception in the presence of exponentially correlated α-μ fading. For example, if the reception is performed by multibranch linear array of antennas, which are equally spaced, then the correlation between individual antenna elements could be treated as exponential.

3.1.2 Constant Correlation Model

If the reception is performed by using multibranch circular symmetric array of antennas, which are closely placed [10], then the correlation between individual antenna elements could be treated as constant. Let us consider the multivariate correlative distribution of n random variables. The correlation model is considered to be constant if the elements of the covariance matrix (correlation coefficients between random variables from multivariate distribution) are formed according to [11]

$$\Sigma_{ij} = \begin{cases} 1, & i = j \\ \rho, & i \neq j \end{cases} \quad 1 \leq i, j \leq n. \tag{3.11}$$

The expression for JPDF, $f_{R_{N_1}, \dots, R_{N_n}}(R_{N_1}, \dots, R_{N_n})$ of n exponentially correlated Nakagami-m random variables, $R_{N_1}, \dots, R_{N_n}$, can be found in the form of [12]

$$f_{R_{N_1}, \dots, R_{N_n}}(R_{N_1}, \dots, R_{N_n})$$

$$= \frac{\left(1 - \sqrt{\rho_N}\right)^m}{\Gamma(m)} \underbrace{\sum_{k_1=0}^{\infty} \dots \sum_{k_n=0}^{\infty}}_{n} \frac{2^n \Gamma(m + k_1 + \dots + k_n) \rho_N^{(k_1 + \dots + k_n)/2}}{\Gamma(m + k_1) \dots \Gamma(m + k_n) k_1! \dots k_n!}$$

$$\times \left(\frac{1}{1 + (n-1)\sqrt{\rho_N}}\right)^{m + k_1 + \dots + k_n} \times \left(\frac{m}{\Omega_1 \left(1 - \sqrt{\rho_N}\right)}\right)^{m + k_1} \dots$$

$$\times \left(\frac{m}{\Omega_n \left(1 - \sqrt{\rho_N}\right)}\right)^{m + k_n} \times R_{N_1}^{2m + 2k_1 - 1} \dots R_{N_n}^{2m + 2k_n - 1}$$

$$\times \exp\left(-\frac{m R_{N_1}^2}{\Omega_1 \left(1 - \sqrt{\rho_N}\right)}\right) \dots \exp\left(-\frac{m R_{N_n}^2}{\Omega_n \left(1 - \sqrt{\rho_N}\right)}\right). \tag{3.12}$$

Substituting (3.12) and (3.7) into (3.6), and with respect to the parameter transformation already explained in [13],

$$\Omega_1 = \hat{R}_1^{\alpha_1}, \dots, \Omega_n = \hat{R}_n^{\alpha_1}, \tag{3.13}$$

an expression for the JPDF of constantly correlated α-μ random variables is obtained in the form of

$$f_{R_1,\ldots,R_n}(R_1,\ldots,R_n)$$

$$= \frac{\left(1-\sqrt{\rho_{\alpha-\mu}}\right)^{\mu}}{\Gamma(\mu)} \sum_{k_1,\ldots,k_n=0}^{\infty} \frac{\Gamma\left(\mu_d + k_1 + \cdots + k_n\right)\rho_{\alpha-\mu}^{(k_1+\cdots+k_n)/2}}{\left(1-\sqrt{\rho_{\alpha-\mu}}\right)^{n\mu+k_1+\cdots+k_n}}$$

$$\times \mu^{n\mu_d + k_1 + \ldots + k_n} \left(\frac{1}{1+(n-1)\sqrt{\rho_{\alpha-\mu}}}\right)^{\mu+k_1+\cdots+k_n}$$

$$\times \prod_{i=1}^{n} \frac{\alpha_i}{\Gamma(\mu+k_i)k_i!\hat{R}_i^{\alpha_i(\mu+k_i)}} R_i^{\alpha_i(\mu+k_i)-1} \exp\left(-\frac{\mu R_i^{\alpha_i}}{\hat{R}_i^{\alpha_i}(1-\sqrt{\rho_{\alpha-\mu}})}\right).$$

$$(3.14)$$

3.1.3 General Correlation Model

At the end of this subsection, the JPDF of n α-μ random variables will be presented, observing the arbitrary correlation between RVs, given with the general correlation model, expressed as [6]

$$f_{R_1,\ldots R_n}(R_1,\ldots R_n) = \prod_{i=1}^{n} f_{R_i}(R_i) \sum_{k=0}^{\infty} \frac{(\mu)_k}{k!}$$

$$\times \left\{ \sum_{i<j} C_{ij} \frac{L\left((\mu/\hat{R}_i^{\alpha_i})R_i^{\alpha_i},\mu\right)}{\mu} \frac{L\left((\mu/\hat{R}_i^{\alpha_j})\hat{R}_i^{\alpha_j},\mu\right)}{\mu} + \cdots \right.$$

$$\left. + C_{123\ldots n} \frac{L\left((\mu/\hat{R}_i^{\alpha_i})R_i^{\alpha_i},\mu\right)}{\mu} \frac{L\left((\mu/\hat{R}_i^{\alpha_n})R_n^{\alpha_n},\mu\right)}{\mu} \right\}^{k},$$

$$(3.15)$$

where

$L_r(x,p)$ is the generalized Laguerre polynomial given by [14]
$(a)_n$ is the Pochhammer symbol [8]

The Krishnamoorthy's coefficients, C_{ij}, from the previous relation, are defined in [15] and they are a function of correlation coefficients:

$$C_{123\ldots n} = (-1)^n \begin{vmatrix} 0 & \rho_{12} & \cdots & \rho_{1n} \\ \rho_{21} & 0 & \cdots & \rho_{2n} \\ \vdots & \vdots & \cdots & \vdots \\ \rho_{n1} & \rho_{n2} & \cdots & 0 \end{vmatrix}_{n \times n}. \tag{3.16}$$

Expressions presented previously have a wide range of applications and represent an excellent basis for further performance analysis of diversity reception, which will be discussed in the following sections.

3.2 Bivariate Rician Distribution

Since dual-branch diversity reception in a Rician-fading environment, in the presence of cochannel interference, will be discussed in Chapter 6, we present here the expression for bivariate correlative Rician distribution in the form of [16]

$$f_{R_1, R_2}(r_1, r_2)$$

$$= \frac{r_1 r_2 (K_d + 1)^2}{\beta_d^2 (1 - r^2)} \exp\left(-\frac{(r_1^2 + r_2^2)(K_d + 1) + 4K_d \beta_d (1 - r)}{2\beta_d (1 - r^2)}\right)$$

$$\times \sum_{k=0}^{\infty} \varepsilon_k I_k \left(\frac{r_1 r_2 r (K_d + 1)}{\beta_d (1 - r^2)}\right) I_k \left(\frac{r_1}{(1 + r)} \sqrt{\frac{2K_d (K_d + 1)}{\beta_d}}\right)$$

$$\times I_k \left(\frac{r_2}{(1 + r)} \sqrt{\frac{2K_d (K_d + 1)}{\beta_d}}\right), \tag{3.17}$$

where

β_d denotes the average power of signal envelopes, defined as $\beta_d = \overline{R_1^2}/2 = \overline{R_2^2}/2$

K_d, known as the Rice factor, defines the ratio of signal power in the dominant components of desired signal over the scattered power

r is the correlation coefficient

constant ε_k is defined as $\varepsilon_k = 1$ $(k = 0)$, and $\varepsilon_k = 2$ $(k \neq 0)$

3.3 Bivariate Hoyt Distribution

Expressions for the JPDF of a bivariate Hoyt (Nakagami-q) process with an arbitrary correlation pattern in a nonstationary environment have been derived recently [17]. A solution to a long-standing unsolved problem is presented in the following form:

$$
f_{R_1,R_2}(r_1,r_2) = \sum_{k=0}^{\infty} \sum_{k_1=0}^{k} \sum_{k_2=0}^{k_1} \sum_{k_3=0}^{k_2} \sum_{k_4=0}^{k_3}
$$

$$
\times \frac{(1/2)_k \, \delta_1^{2(k-k_1)} \delta_4^{2(k_1-k_2)} \delta_3^{2(k_2-k_3)} \delta_2^{2(k_3-k_4)} (-1)^{k_4} (\delta_1\delta_2 - \delta_3\delta_4)^{2k_4}}{(k-k_1)!(k_1-k_2)!(k_2-k_3)!(k_3-k_4)!k_4!}
$$

$$
\times \sum_{j_1=0}^{k-k_1+k_2-k_3+k_4} \sum_{j_2=0}^{k_1-k_2+k_3} \frac{(-1)^{j_1+j_2}}{\eta_1^{j_1+1/2}} \binom{k-k_1+k_2-k_3+k_4}{j_1}\binom{k_1-k_2+k_3}{j_2}
$$

$$
\times \sum_{j_3=0}^{\infty} \frac{(j_1+1/2)_{j_3}}{j_3!} \left(\frac{\eta_1-1}{\eta_1}\right)^{j_3} P\left(j_1+j_2+j_3+1, \frac{1+\eta_1}{2\Omega_1}r_1^2\right)
$$

$$
\times \sum_{j_4=0}^{k-k_2+k_4} \sum_{j_5=0}^{k_2} \frac{(-1)^{j_4+j_5}}{\eta_2^{j_4+1/2}} \times \binom{k-k_2+k_4}{j_4}\binom{k_2}{j_5}
$$

$$
\times \sum_{j_6=0}^{\infty} \frac{(j_4+1/2)_{j_6}}{j_6!} \left(\frac{\eta_2-1}{\eta_2}\right)^{j_6} P\left(j_4+j_5+j_6+1, \frac{1+\eta_2}{2\Omega_2}r_2^2\right),
$$

$$
\tag{3.18}
$$

where

the correlation coefficients δ_1, δ_2, δ_3, and δ_4 depend on several parameters, such as the distance between reception points and the frequency difference between transmitted signals, among others [18,19]

$P(a, x)$ is the regularized incomplete gamma function [8, Eq. (6.5.1)]

$\Omega_1 = E(r_1^2)$

$\Omega_2 = E(r_2^2)$

η_1 and η_2 define scattered wave power ratios between the in-phase and the quadrature component of each cluster

3.4 Bivariate Generalized K Distribution

Since dual-branch diversity reception in a generalized K fading environment, in the presence of CCI, will be discussed in Chapter 6, we present here the expression for bivariate correlative generalized K distribution in the form of [20]

$$f_{R_1,R_2}\left(R_1,R_2\right) = \frac{16}{\Gamma(m_d)\Gamma(k_d)} \sum_{i,j=0}^{+\infty} \frac{m_d^{\xi_d}\, \rho_{Nd}^i \rho_{Gd}^j}{\Gamma(m_d+i)\Gamma(k_d+j)}$$

$$\left(R_1/\sqrt{\Omega_{d1}}\right)^{\xi_d} K_{\psi_d}\left(2\sqrt{m_d/\sigma_{d1}}\,R_1\right)\left(R_2/\sqrt{\Omega_{d2}}\right)^{\xi_d} K_{\psi_d}$$

$$\times \frac{\left(2\sqrt{(m_d/\sigma_{d2})}R_2\right)}{i!\,j!\left(1-\rho_{Nd}\right)^{k_d+i+j}\left(1-\rho_{Gd}\right)^{m_d+i+j} R_1 R_2}, \qquad (3.19)$$

with

$$\begin{aligned}
\xi_d &= k_d+j+m_d+i, \\
\psi_d &= k_d+j-m_d-i, \\
\sigma_{dl} &= \left(1-\rho_{Nd}\right)\left(1-\rho_{Gd}\right)\Omega_{dl}, \quad l=1,2,
\end{aligned} \qquad (3.20)$$

where

$m_d \geq 1/2$ represents the Nakagami-m shaping parameters

$k_d > 0$ denotes shadowing shaping parameters of the signal, which approximate several shadowing conditions, from severe shadowing ($k_d \to 0$) to no shadowing ($k_d \to \infty$)

ρ_{Nd} is the power correlation coefficient between instantaneous powers of Nakagami-m fading processes

ρ_{Gd} is the correlation coefficient between average fading powers of signals

Ω_{dl} denotes average powers of signals affected by the previously mentioned composite fading/shadowing

K_v (.) denotes the modified Bessel function of the second kind and order v [8, Eq. (8.407/1)]

References

1. Simon, M. K. and Alouini, M. S. (2005). *Digital Communications over Fading Channels*, 2nd edn. Wiley, New York.

2. Pierce, J. N. and Stein, S. (1960). Multiple diversity with non independent fading. *Proceedings of the Institute of Radio Engineers*, 48, 89–104.

3. Aalo, V. A. (1995). Performance of maximal-ratio diversity systems in a correlated Nakagami fading environment. *IEEE Transactions on Communications*, 43(8), 2360–2369.

4. Turkmani, A. M. D. et al. (1995). An experimental evaluation of the performance of two-branch space and polarization diversity schemes at 1800 MHz. *IEEE Transactions on Vehicular Technology*, 44(2), 318–326.

5. Nakagami, M. (1960). The *m*-distribution—A general formula of intensity distribution of rapid fading. *Statistical Methods in Radio Wave Propagation*. Pergamon Press, Oxford, U.K., pp. 3–36.

6. De Souza, R. A. A., Fraidenraich, G., and Yacoub, M. D. (2006). On the multivariate α-μ distribution with arbitrary correlation. *Proceedings of VI International Telecommunications Symposium*, Fortaleza, Ceara, Brazil, pp. 38–41.

7. Karagiannidis, G. K., Zogas, D. A., and Kotsopoulos, S.A. (2003). On the multivariate Nakagami-m distribution with exponential correlation. *IEEE Transactions on Communications*, 51(8), 1240–1244.

8. Gradshteyn, I. and Ryzhik, I. (1980). *Tables of Integrals, Series, and Products*. Academic Press, New York.

9. Proakis, J. G. (2001). *Digital Communications*. McGraw-Hill, New York, 2001.

10. Vanghn, R. G. and Anderson, J. B. (1987). Antenna diversity in mobile communications. *IEEE Transactions on Vehicular Technology*, 36(4), 149–172.

11. Karagiannidis, G. K., Zogas, D. A., and Kotsopoulos, S. A. (2003). An efficient approach to multivariate Nakagami-m distribution using Green's matrix approximation. *IEEE Transactions on Wireless Communications*, 2(5), 883–889.

12. Reig, J. (2007). Multivariate Nakagami-*m* distribution with constant correlation model. *International Journal of Electronics and Communications (AEUE)*, 63(1), 46–51.

13. Yacoub, M. D. (2002). The α-μ distribution: A general fading distribution. *Proceeding of the 13th International Symposium on Personal, Indoor, and Mobile Radio Communications*, Lisaboa, Portugal, Vol. 2, pp. 629–633.

14. Wolfram Research, Inc. http://mathworld.wolfram.com/LaguerrePoly nomial.html, accessed June 2012.

15. Krishnamoorthy, S. A. and Parthasarathy, M. (1951). A multivariate gamma-type distribution. *Annals of American Mathematics*, 22(4), 549–557.

16. Bandjur, D. V., Stefanovic, M. C., and Bandjur, M. V. (2008). Performance analysis of SSC diversity receivers over correlated Ricean fading channels in the presence of co-channel interference. *Electronic Letters*, 44(9), 587–588.

17. De Souza, R. A. A. and Yacoub, M. D. (2009). Bivariate Nakagami-q (Hoyt) distribution. *IEEE Transactions on Communications*, 60(3), 714–723.

18. Jakes, W. (1997). *Microwave Mobile Communications*. Wiley, New York.

19. Adachi, F., Feeney, M. T., and Parsons, J. D. (1988). Effects of correlated fading on level crossing rates and average fade durations with predetection diversity reception. *IEE Proceedings F, Radar Signal Process*, 135(1), 11–17.
20. Bithas, P., Sagias, N., and Mathiopoulos, P. (2009). The bivariate generalized-K (KG) distribution and its application to diversity receivers. *IEEE Trans. Commun.*, 57(9), 2655–2662.

4

PERFORMANCES OF DIGITAL RECEIVERS

4.1 System Performance Measures

In this chapter, several measures of performances related to wireless communication system design will be defined, and mathematical methods for their evaluation will be presented. Capitalizing on the formal definitions of these measures, depending on evaluation complexity and analytical tractability of the obtained mathematical form, some of them will be presented in subsequent chapters for observed communication scenarios.

4.1.1 *Average Signal-to-Noise Ratio*

A very important parameter for describing and measuring the reception sensitivity of any communication system is signal-to-noise ratio (SNR). However, communication channel noise is not often the ultimate drawback. When considering a system subjected to the influence of fading, modeled by some statistical distribution, a more appropriate measure for describing receiver sensitivity is average SNR, with the received SNR statistically averaged over the probability density function (PDF) of the fading [1].

This measure can be determined from

$$\bar{x} \triangleq \int_0^\infty x f_X(x)\,dx, \qquad (4.1)$$

with $f_X(x)$ being the PDF of x, the instantaneous value of the SNR random process. In addition, if the moment-generating function (MGF) of instantaneous SNR is known in advance, the average SNR value can be delivered from

$$\bar{x} \triangleq \frac{\partial M_x(s)}{\partial s}\bigg|_{s=0}. \tag{4.2}$$

Often, this approach is used for determining the average SNR at the output of the receiver when the maximal-ratio combining (MRC) space diversity technique is applied at the reception and no correlation between the channels is assumed. In such a case, as described later, the output SNR is formed as a sum of the individual channel SNRs, x_i, as

$$x = \sum_{i=1}^{N} x_i, \tag{4.3}$$

with N denoting the number of combining channels. Since channels are assumed to be independent, the MGF can be formed as the product of the individual channel MGFs, $Mx_i(s)$, as

$$M_x(s) = \prod_{i=1}^{N} M_{x_i}(s). \tag{4.4}$$

This property simplifies the evaluation of the average SNR by differentiating the previous expression with respect to s, instead of forming joint PDF of SNRs by convoluting corresponding SNRs.

4.1.2 Outage Probability

Outage probability (OP, P_{out}) is another accepted performance measure for describing diversity systems operating in fading environments. This measure is crucial in helping the wireless communications system designers to meet the quality of service (QoS) and grade of service (GoS) demands. OP is defined in terms of received signal statistics, as the probability that the received SNR, x, falls below a certain given outage threshold, x_{th}, also known as a protection ratio. If the CDF of x is known in advance, OP can be calculated from

$$P_{out} = \int_{0}^{x_{th}} f_x(t)dt = F_x(x_{th}). \tag{4.5}$$

Knowing the properties of MGF, and the connection between the MGF and PDF explained in Chapter 1, based on the previous relation, it is evident that OP can also be determined as the inverse Laplace transform of the ratio $M_x(-s)/s$ evaluated at $x = x_{th}$, that is,

$$P_{out} = \frac{1}{2\pi j} \int_{\sigma-j\infty}^{\sigma-j\infty} \frac{M_x(-s)}{s} \exp(sx_{th}) ds; \qquad (4.6)$$

with σ chosen in the region of convergence of the integral in the complex s plane.

OP is very useful in wireless communication systems design especially for the cases where cochannel interference (CCI) is present. In the interference-limited environment, OP is defined as the probability that the output signal-to-interference ratio (SIR) falls below a given outage threshold.

Outage is closely related to the reuse distance and the coverage (service) area of a cellular system [2]. The reuse distance is the minimum distance between any two cochannel base stations, which ensures a worst-case OP no larger than the required value. Reuse distance, also known as the CCI reduction factor, can be calculated as a function of OP and is used for estimating the spectral efficiency [3] and determining the reuse pattern. Another system parameter, which can be calculated based on OP, is the coverage area, the area within which OP is guaranteed to be less than a given threshold.

4.1.3 Average Symbol Error Probability

Performance measure, which in the best way describes the nature of the wireless communication system behavior, is the average symbol error probability (ASEP) or alternatively average symbol error rate (ASER). If the number of bits per symbol is equal to 2, then this measure is equivalent to the measure known as average bit error probability (ABEP) or alternatively average bit error rate (ABER). Otherwise, if we want to obtain ABER values, the signal energy per symbol should be converted into signal energy per bit. ASEP values are obtained capitalizing on conditional SEP relations, which are conditioned over fading statistics, which impairs communication. If conditional SEP is denoted with $P_s(e|x)$, then by averaging over SNR, ASEP can be obtained as

$$\bar{P_e} = \int_0^\infty P_s(e|x) f_x(x) dx. \tag{4.7}$$

Conditional SEP is the function of the instantaneous SNR, and functional dependency is determined by the type of modulation scheme performed. The instantaneous SEP of several common modulation schemes [4] is given in Table 4.1, with *erfc* (*x*) being the complementary error function [8, Eq. (8.250.4)].

Modulation schemes are classified according to the carrier property being modulated, the signal amplitude, signal frequency, or signal phase. Another classification is based on the number of levels assigned to a modulated carrier. Finally, modulation techniques can be divided according to the quantity of necessary carrier-phase information needed to be extracted in the detection process (coherent, noncoherent modulations, etc.).

Another approach for determining ASEP performances of a wireless communication system exposed to fading influence is the MGF-based approach, which eliminates possible integration difficulties.

By using some alternate representations of functions in conditional SEP expressions after some mathematical transformations, $P_s(e|x)$ can be written in the following form:

$$P_s(e|x) = \sum \int_{\theta_1}^{\theta_2} h(\theta) \exp\left(-g(\theta)x\right) d\theta, \tag{4.8}$$

with the number of terms in the sum, parameters θ_1 and θ_2 and functions $g(\theta)$ and $h(\theta)$, depending on observed conditional SEP functions (applied modulation technique). Now, after substituting (4.8) into (4.7), we obtain the following transformation:

$$\bar{P_e} = \sum \int_{\theta_1}^{\theta_2} h(\theta) \underbrace{\int_0^\infty f_x(x) \exp\left(-g(\theta)x\right) dx}_{M_x(-g(\theta))} d\theta$$

$$= \sum \int_{\theta_1}^{\theta_2} h(\theta) M_x\left(-g(\theta)\right) d\theta, \tag{4.9}$$

Table 4.1   Instantaneous SEP for Some Modulation Schemes

MODULATION SCHEME	CONDITIONAL SEP $P_s(e|x)$
Coherent binary signaling	
Coherent phase-shift keying (CPSK)	$0.5\,erfc\sqrt{x}$
Coherent detection of differentially encoded PSK	$erfc\sqrt{x} - 0.5\,erfc^2\sqrt{x}$
Coherent frequency-shift keying (CFSK)	$0.5\,erfc\sqrt{x/2}$
Noncoherent binary signaling	
Differential phase-shift keying (DPSK)	$0.5\exp(-x)$
Noncoherent frequency-shift keying (NCFSK)	$0.5\exp(-x/2)$
Quadrature signaling	
Quadrature phase-shift keying (QPSK)	$erfc\sqrt{x} - 0.25\,erfc^2\sqrt{x}$
Minimum shift keying (MSK)	$erfc\sqrt{x} - 0.25\,erfc^2\sqrt{x}$
$\pi/4$ Differential quaternary PSK with Gray coding [5]	$\dfrac{1}{2\pi}\displaystyle\int_0^\pi \dfrac{\exp(-x(2-\sqrt{2}\cos\theta))}{\sqrt{2}-\cos\theta}\,d\theta$
Multilevel signaling	
Square quadrature amplitude modulation (SQAM)	$2q\,erfc\sqrt{px} - q^2 erfc^2\sqrt{px};$ $q = 1 - \dfrac{1}{\sqrt{M}};\ p = 1.5\log_2\dfrac{M}{M-1};$
Multiple phase-shift keying (MPSK)	$\dfrac{1}{\pi}\displaystyle\int_0^{\pi-\pi/M}\exp\left(\dfrac{-x\sin^2(\pi/M)\log_2 M}{\sin\theta}\right)d\theta$
M-ary differential phase-shift keying (MDPSK) [6]	$\dfrac{\sin(\pi/M)}{\pi}\displaystyle\int_0^{\pi/2}\dfrac{\exp(-x\log_2 M[1-\cos(\pi/M)\cos\theta])}{1-\cos(\pi/M)\cos\theta}\,d\theta$ or $\dfrac{1}{\pi}\displaystyle\int_0^{\pi-\pi/M}\exp\left(\dfrac{-x\sin^2(\pi/M)\log_2 M}{1+\cos(\pi/M)\cos\theta}\right)d\theta$
Two-dimensional M-ary constellations [7]	$\dfrac{1}{2\pi}\displaystyle\sum_{k=1}^{N}P_r(S_k)\int_0^{\eta_k}\exp\left(\dfrac{-x\alpha\sin^2(\psi_k)}{\sin^2(\theta+\psi_k)}\right)d\theta$

Where *N* is the number of signal points and $P_r(S_k)$ is the a priori probability that the *k*th signal point is transmitted

Table 4.2   ASEP for Some Modulation Schemes Defined through the MGF of SNR

MODULATION SCHEME	ASEP
Binary phase-shift keying (BPSK)	$\dfrac{1}{\pi}\displaystyle\int_0^{\pi/2} M_x\left(-\dfrac{1}{\sin^2\theta}\right)d\theta$
Orthogonal binary frequency-shift keying	$\dfrac{1}{\pi}\displaystyle\int_0^{\pi/2} M_x\left(-\dfrac{1}{2\sin^2\theta}\right)d\theta$
Binary frequency-shift keying with minimum correlation [9]	$\dfrac{1}{\pi}\displaystyle\int_0^{\pi/2} M_x\left(-\dfrac{0.715}{2\sin^2\theta}\right)d\theta$
Multiple phase-shift keying (MPSK)	$\dfrac{1}{\pi}\displaystyle\int_0^{\pi-\pi/M} M_x\left(\dfrac{\sin^2(\pi/M)}{\sin^2\theta}\right)d\theta$
Multiple differential phase-shift keying (MDPSK)	$\dfrac{2}{\pi}\displaystyle\int_0^{(M-1)\pi/2M} M_x\left(\dfrac{\sin^2(\pi/M)}{2\cos^2(\pi/2M)-2\cos(\pi/M)\sin^2\theta}\right)d\theta$
M-ary quadrature amplitude modulation (M-QAM)	$\dfrac{4}{\pi}\left(1-\dfrac{1}{\sqrt{M}}\right)\displaystyle\int_0^{\pi/2} M_x\left(\dfrac{3}{2(M-1)\sin^2\theta}\right)d\theta$
	$-\dfrac{4}{\pi}\left(1-\dfrac{1}{\sqrt{M}}\right)\displaystyle\int_0^{\pi/4} M_x\left(\dfrac{3}{2(M-1)\sin^2\theta}\right)d\theta$

so the final result is expressed in the form of MGF of x. In Table 4.2, the ASEP expressions are presented for various modulation techniques defined through the MGF of SNR of the fading process.

In addition, the ASEP expression of any two-dimensional (2D) amplitude/phase linear modulation can be found as a weighted sum of MGFs [10]:

$$\bar{P}_e = \sum_{i=1}^{L} \frac{w_i}{2\pi} \int_0^{\theta} M_x\left(-\frac{a_i \sin^2\psi_i}{\sin^2(\theta+\psi_i)}\right) d\theta_i, \qquad (4.10)$$

where

L is the number of decision regions

w_i is the a priori probability of the symbol to which the region corresponds

a_i is a normalization factor

the angles θ_i and ψ_i are determined by the geometry of the 2D constellations and the resulting decision regions

ASEP expressions for various differentially coherent and noncoherent signals could also be found in [11].

4.1.4 Amount of Fading

The performance measure describing fading channel severity, obtained by first and second central moments of received SNR, has been introduced in the literature as the amount of fading (AoF) [12]. This easy, analytically traceable measure was generalized for describing the behavior of diversity combining at the reception [13]. Through this measure, it has been shown that diversity combining at the reception reduces the relative variance of the received envelope that cannot be achieved only by increasing the transmitter power.

If total instantaneous SNR at the receiver is denoted with x, AoF can be determined according to

$$AoF = \frac{Var(x)}{\left(E(x)\right)^2} = \frac{E(x^2) - \left(E(x)\right)^2}{\left(E(x)\right)^2}, \tag{4.11}$$

where $Var(x)$ and $E(x)$ denote variance and mathematical expectation operators. The MGF approach for determining AoF is based on the definition of MGF of x, and it stands

$$AoF = \frac{\partial^2 M_x(s)/\partial s^2 \big|_{s=0} - \left(\partial M_x(s)/\partial s \big|_{s-0}\right)^2}{\left(\partial M_x(s)/\partial s \big|_{s=0}\right)^2}. \tag{4.12}$$

AoF should be observed as a reception measure of the performance of the entire system, either with or without diversity reception applied.

4.1.5 Level Crossing Rate

Exposed to the influence of various drawbacks that occur during a wireless transmission, the signal at the reception is subjected to heavy statistical fluctuations of even 30 dB. As known, a decrease in the received signal quality causes an increase in the ASER. In order to improve the ASER performance, the communication system designer must be aware how often the receiver signal crosses the given signal level per time unit and for how long an average received signal stays below the observed signal level. Time-dependent measures (second-order statistical parameters) that describe those two properties are level crossing rate (LCR) and average fade duration (AFD).

LCR, introduced in [14], is defined as the rate at which a random process crosses level z in a positive or a negative direction. Mathematically;

$$N_Z(z) = \int_0^\infty \dot{z} f_{z,\dot{z}}(z, \dot{z}) d\dot{z} \qquad (4.13)$$

with z denoting the received signal envelope random process, and $\dot{z}$ its time derivative at the same time instant, with their JPDF being denoted as $f_{z,\dot{z}}(z, \dot{z})$.

4.1.6 Average Fade Duration

The received signal fade duration defines the expected number of signaling bits that will be lost during the fade and is a function of several transmission factors, among which an important factor is the moving speed of the mobile-radio/satellite vehicle.

AFD is defined as the average time over which the signal envelope ratio remains below a specified level after crossing that level in a downward direction and is determined as

$$T_Z(z) = \frac{F_z(z \le Z)}{N_Z(z)}. \qquad (4.14)$$

AFD and LCR values are related to the criterion used to assess the error probability of packets of distinct length and to determine parameters of equivalent channel, modeled by a Markov chain with a defined number of states [15].

4.2 Space Diversity Combining

Diversity combining represents a concept of increasing system performances at the reception by combining two or more replicas of information bearing signal. The main idea is to utilize the low probability of deep fade concurrence in all signal replicas, in order to decrease the probability of error and outage occurrence.

Obtaining multiple replicas of the same information bearing signal could be performed by following various procedures. Some of them are channel coding in combination with limited interleaving—*time diversity*; transmitting the same narrowband signal at different carrier frequencies, where the carriers are separated by the coherence bandwidth of the channel—*frequency diversity* (*path diversity*, i.e., frequency hopping in GSM, multicarrier systems, spread spectrum systems, etc.); using two transmit antennas/two receive antennas with different polarization—*polarization diversity* [16,17]. However, all of the described techniques have some appliance disadvantages compared to *space diversity* [1].

Here, by usage of multiple receive antennas (antenna array), with separated antennas on diversity terminal, independent replicas of information bearing signal are realized, without an increase in signal bandwidth or transmitted signal power. Two criteria are necessary to obtain a high degree of improvement from a space diversity system. The first criterion is that the fading in individual branches should have low cross-correlation. Correlation arises between branches when a diversity system is applied on small terminals with multiple antennas. If the correlation is too high, then deep fades in the branches will occur simultaneously. So, the separation distance between antennas must guarantee approximately independent amplitudes at diversity branches (i.e., about 0.38 wavelength in uniform isotropic scattering environment, larger separation for directional antennas, etc.). The second criterion is that the mean power available from each branch should be almost equal. If reception branches have low correlation, but have very different mean power, then the signal in a weaker branch may not be useful even though it is less faded (below its mean) than the other branches. Independent signal replicas are then combined at the reception (see Figure 9.1 from [1]) by using several methods that differ in complexity and overall performance, and the resultant signals are then passed through a demodulator. In this book some of those methods will be considered: maximal ratio combining (MRC), equal gain combining (EGC), selection combining (SC), and switch-and-stay combining (SSC).

In order to perform appropriate design of a high-performance wireless system with defined QoS and GoS, taking into account the predetermined complexity constraints, trade-off analysis must be carried out, considering various combinations between applied modulation/ coding/diversity techniques, and a decision should be made based on precise quantitative performance evaluation of observed combinations. Further in the book, we will analyze standard wireless performance measures, defined in this Chapter in the function of various system parameters, such as fading statistics over communication channels, fading severity, average fading power, diversity type applied, diversity order, correlation level between branches, diversity branch imbalance, etc.

4.2.1 Maximal Ratio Combining

The optimal combining technique, regardless of fading statistics, is maximum ratio combining (MRC). This combining technique involves cophasing of the useful signal in all branches, multiplication of the received signal in each branch by the estimated envelope of that particular signal, and summing of the received signals from all antennas [1]. By cophasing, all the random phase fluctuations of the signal that emerged during the transmission are eliminated. For this process, it is necessary to estimate the phase of the received signal, so this technique requires the entire amount of the channel state information of the received signal and a separate receiver chain for each branch of the diversity system, which increases the complexity and expense of the system. Because fading amplitudes over all channels are known, MRC could be used in conjunction with unequal energy signals, such as M-QAM or any other amplitude/phase modulations and coherent modulation schemes, but it is not practical for noncoherent transmission techniques.

As explained with MRC combining, the received signals are cophased, each signal is amplified appropriately for optimal combining, and the resulting signals are added so the envelope at the MRC output R is given with

$$R = \sqrt{\sum_{i=1}^{M} R_i^2}.$$ (4.15)

The PDF of the envelope random process at the output of the MRC can be written as [18,19]

$$f_R(r) = \int_0^{r} \int_0^{\sqrt{r^2 - r_N^2}} \cdots \int_0^{\sqrt{r^2 - \sum_{i=3}^N r_i^2}} \frac{r}{\sqrt{r^2 - \sum_{i=2}^N r_i^2}}$$

$$\times f_{R_1,\ldots,R_N}\left(\sqrt{r^2 - \sum_{i=2}^N r_i^2}, r_2 \ldots r_N\right) dr_2 \ldots dr_N, \qquad (4.16)$$

with $f_{R_1,\ldots,R_N}\left(r_1, r_2, \ldots, r_N\right)$ being the multivariate JPDF of random processes at input branches. If reception is performed with uncorrelated input branches, then the previous expression reduces to

$$f_R(r) = \int_0^{r} \int_0^{\sqrt{r^2 - r_N^2}} \cdots \int_0^{\sqrt{r^2 - \sum_{i=3}^N r_i^2}} \frac{r}{\sqrt{r^2 - \sum_{i-2}^N r_i^2}}$$

$$\times f_{r_1,\ldots,r_N}\left(\sqrt{r^2 - \sum_{i=2}^N r_i^2}\right) \prod_{i=2}^N f_{R_i}(r_i) dr_2 \ldots dr_N. \qquad (4.17)$$

Similarly, the CDF of the MRC output can be determined according to

$$F_R(r) = \int_0^{r} \int_0^{\sqrt{r^2 - r_N^2}} \cdots \int_0^{\sqrt{r^2 - \sum_{i=3}^N r_i^2}} \int_0^{\sqrt{r^2 - \sum_{i=2}^N r_i^2}} f_{R_1,\ldots,R_N}\left(r_1, r_2, \ldots, r_N\right) dr_1 dr_2 \ldots dr_N.$$

$$(4.18)$$

Capitalizing on (4.16), by using (2.3) and (2.6), the MGF of the output SNR could be obtained. An especially interesting case is when channels (diversity branches) are assumed to be independent and then MGF can be formed as the product of the individual channel MGFs, $Mx_i(s)$, as

$$M_x(s) = \prod_{i=1}^N M_{x_i}(s), \qquad (4.19)$$

where $Mx_i(s)$ can be determined according to (2.6).

4.2.2 Equal Gain Combining

Equal Gain Combining (EGC) is a suboptimal space diversity reception technique. This technique has reduced complexity compared to the MRC receiving technique (see Figure 9.2 from [1]), since it does not require estimation of the channel (path) fading amplitudes, but only requires channel carrier phase estimation, for equaling weights applied to each branch in combination sum (amplitudes of complex weights should have value 1, while phases should be equal to negatives of the estimated carrier phase). In cases where coherent detection could not be performed, or in cases where noncoherent detection should be carried out, the MRC technique could not be applied. Therefore, in such cases, reception is carried out by using the postdetection EGC reception technique (i.e., DPSK and FSK modulation cases).

As already explained, the EGC technique processes all diversity branches and sums received equally weighted replicas from each one to produce output statistics, so the envelope at the N-branch EGC output R is given by [18]

$$R = \frac{1}{\sqrt{N}} \sum_{i=1}^{N} R_i.$$ (4.20)

The PDF of the envelope random process at the output of EGC can be written as [18,19]

$$f_R(r) = \sqrt{N} \int_0^{\sqrt{N}r} \int_0^{\sqrt{N}r-r_N} \cdots$$

$$\times \int_0^{\sqrt{N}r-\sum_{i=3}^{N} r_i} f_{R_1,\dots,R_N}\left(r\sqrt{N} - \sum_{i=2}^{N} r_i, r_2 \dots r_N\right) dr_2 \dots dr_N,$$ (4.21)

with $f_{R_1,\dots,R_N}\left(r_1, r_2, \dots, r_N\right)$ being the multivariate JPDF of random processes at the input branches. If reception is performed with uncorrelated input branches, then the previous expression reduces to

$$f_R(r) = \sqrt{N} \int_0^{\sqrt{N}r} \int_0^{\sqrt{N}r - r_N} \cdots$$

$$\times \int_0^{\sqrt{N}r - \sum_{i=3}^N r_i} f_{R_1}\left(r\sqrt{N} - \sum_{i=2}^N r_i \right) \prod_{i=2}^N f_{R_i}(r_i)\, dr_2 \cdots dr_N.$$

$$(4.22)$$

Similarly, the CDF of the EGC output can be determined according to

$$F_R(r) = \int_0^{\sqrt{N}r} \int_0^{\sqrt{N}r - r_N} \cdots$$

$$\times \int_0^{\sqrt{N}r - \sum_{i=3}^N r_i} \int_0^{\sqrt{N}r - \sum_{i=2}^N r_i} f_{R_1,\dots,R_N}\left(r_1, r_2 \dots r_N \right) dr_1 dr_2 \dots dr_{N-1} dr_N.$$

$$(4.23)$$

Based on expression (4.21), after using (2.3) and (2.6), the MGF of the output SNR could be efficiently evaluated.

4.2.3 Selection Combining

As already explained, the MRC and EGC combining techniques require a separate receiver chain for each branch of the diversity system, which increases its complexity. The selection combining (SC) technique receiver is much simpler for practical realization (trade-off with obtained reception performances), in opposition to these combining techniques, since it processes one of the received signal replicas. However, this combining technique still requires the simultaneous and continuous monitoring of all the channels. In general, SC, assuming that noise power is equally distributed over branches, selects the branch with the highest SNR, that is, the branch with the strongest signal. Similar to the previous, there is a type of SC that chooses the branch with the highest signal and noise sum [20]. In fading environments as in cellular systems where the level of the CCI is sufficiently high as compared to the thermal noise, SC selects the branch with the highest SIR (SIR-based selection diversity) [21]. This type of SC in which the branch with the highest SIR is selected can be measured in real time both in base stations and in mobile stations using specific SIR estimators as well as those for both analog and digital wireless systems (e.g., GSM, IS-54) [22,23]. Since cophasing of multiple branches is not required, SC can be used in conjunction with either coherent or differential modulation.

As mentioned, the SC diversity receiver selects and outputs the branch with the highest instantaneous signal envelope value, that is,

$$R = R_{out} = \max(R_1, R_2, \ldots, R_N). \tag{4.24}$$

The PDF of the envelope random process at the output of SC can be written as [24]

$$
\begin{aligned}
f_R(r) = & \underbrace{\int_0^r \int_0^r \cdots \int_0^r}_{n-1} f_{R_1,\ldots,R_N}\left(r, r_2, \ldots, r_N\right) dr_2 dr_3 \ldots dr_N \\
& + \underbrace{\int_0^r \int_0^r \cdots \int_0^r}_{n-1} f_{R_1,\ldots,R_N}\left(r_1, r, \ldots, r_N\right) dr_1 dr_3 \ldots dr_N \\
& \vdots \\
& + \underbrace{\int_0^r \int_0^r \cdots \int_0^r}_{n-1} f_{R_1,\ldots,R_N}\left(r_1, r_2, \ldots, r\right) dr_1 dr_2 \ldots dr_{N-1},
\end{aligned}
\tag{4.25}
$$

with $f_{R_1,\ldots,R_N}\left(r_1, r_2, \ldots, r_N\right)$ being the multivariate JPDF of random processes at input branches. If reception is performed with uncorrelated input branches, then the previous expression reduces to

$$f_R(r) = \sum_{i=1}^n f_{R_i}(r) \prod_{\substack{j=1 \\ j \neq i}}^n F_{r_i}(r); \quad F_{R_i}(r_i) = \int_0^{r_i} f_{t_i}(t_i) dt_i. \tag{4.26}$$

Similarly, the CDF of the SC output can be determined according to [24]

$$F_R(r) = \underbrace{\int_0^r \int_0^r \cdots \int_0^r}_{n} f_{R_1,R_2,\dots,R_n}(r_1,r_2,\dots,r_n) \, dr_1 dr_2 \cdots dr_n. \qquad (4.27)$$

If reception is performed with uncorrelated input branches, then the previous expression reduces to

$$F_R(r) = \prod_{i=1}^{N} F_{R_i}(r). \qquad (4.28)$$

The MGF of the output SNR could be efficiently evaluated by following the explained procedure using expressions (4.25), (2.3), and (2.6).

4.2.4 Switch-and-Stay Combining

In order to avoid the simultaneous and continuous monitoring of all diversity channels, which is necessary with SC reception, a similar form of switched combining, called switch-and-stay combining (SSC), is often used. This is the least complex space diversity reception technique that can be used in conjunction with various coherent, noncoherent, and differentially coherent modulation schemes. Combining is performed in a way that SSC selects and outputs a particular branch until its SNR drops below a predetermined threshold, when the combiner switches to another branch and stays there regardless of the SNR of that branch [25]. Predetermined threshold selection is a very important design issue in SSC reception design. Selection of very high values of the threshold level would cause very frequent switching between the branches, which would lead to poor reception performances. On the contrary, choosing low values of the threshold level would lead to reception performances that are almost equal to no diversity case since reception would be almost locked to one of the diversity branches, even with very low SNR levels on it. Therefore, the optimal threshold value should be determined, and its determination is usually based on ASEP and/or OP values minimization criterion.

Let us denote the predetermined switching threshold for both the input branches with R_T. The PDF of the SSC output can be presented in the form of [26,27]

$$f_{SSC}(r) = \begin{cases} v_{SSC}(r); & r \leq R_T; \\ v_{SSC}(r) + f_{R_1}(r); & r > R_T; \end{cases}$$

$$v_{SSC}(r) = \int_0^{R_T} f_{R_1,R_2}(r,r_2)\,dr_2.$$

(4.29)

The CDF of the SSC output can be presented as [27]

$$F_{SSC}(r) = \begin{cases} F_{R_1,R_2}(r,R_T) & r \leq R_T; \\ F_{R_1}(r) - F_{R_2}(R_T) + F_{R_1,R_2}(r,R_T) & r > R_T; \end{cases}$$

$$F_{R_1,R_2}(r,R_T) = \int_0^r \int_0^{R_T} f_{R_1,R_2}(R_1,R_2)\,dR_1 dR_2;$$

(4.30)

$$F_{R_1}(r) = \int_0^r f_{R_1}(t)\,dt; \quad F_{R_2}(R_T) = \int_0^{R_T} f_{R_2}(t)\,dt.$$

When we consider the case of SSC reception with uncorrelated branches, (4.30) reduces to [1]

$$F_{SSC}(r) = \begin{cases} \dfrac{F_{R_1}(R_T)F_{R_2}(R_T)}{F_{R_1}(R_T) + F_{R_2}(R_T)}\left(F_{R_1}(r) + F_{R_2}(r)\right) & r \leq R_T; \\[4mm] \dfrac{F_{R_1}(R_T)F_{R_2}(R_T)}{F_{R_1}(R_T) + F_{R_2}(R_T)}\left(F_{R_1}(r) + F_{R_2}(r) - 2\right) \\[4mm] \quad + \dfrac{F_{R_1}(r)F_{R_2}(R_T) + F_{R_1}(R_T)F_{R_2}(r)}{F_{R_1}(R_T) + F_{R_2}(R_T)}; & r > R_T. \end{cases}$$

(4.31)

Further, when the SSC over independent identically distributed branches is observed, (4.31) further reduces to

$$F_{SSC}(r) = \begin{cases} F_R(R_T)F_R(r); & r \leq R_T; \\ F_R(r) - F_R(R_T) + F_R(R_T)F_R(r); & r > R_T. \end{cases}$$

(4.32)

Similarly, the PDF of the SSC output for those two special cases reduces to

$$
p_{SSC}(r) = \begin{cases}
\dfrac{F_{R_1}(R_T)F_{R_2}(R_T)}{F_{R_1}(R_T)+F_{R_2}(R_T)}(f_{R_1}(r)+f_{R_2}(r)); & r \leq R_T; \\[4mm]
\dfrac{F_{R_1}(R_T)F_{R_2}(R_T)}{F_{R_1}(R_T)+F_{R_2}(R_T)}(f_{R_1}(r)+f_{R_2}(r)) \\[3mm]
+\dfrac{f_{R_1}(r)F_{R_2}(R_T)+F_{R_1}(R_T)f_{R_2}(r)}{F_{R_1}(R_T)+F_{R_2}(R_T)}; & r > R_T;
\end{cases}
$$

$$(4.33)$$

and

$$
F_{SSC}(r) = \begin{cases}
F_R(R_T)f_R(r); & r \leq R_T; \\
(1+F_R(R_T))f_R(r); & r > R_T.
\end{cases}
\qquad (4.34)
$$

respectively. For all these cases discussed, the corresponding MGFs of SNR could be obtained based on transforming the PDFs from (4.29), (4.33), and (4.34), with respect to (2.3) and further by substituting into (2.6).

4.3 Macrodiversity Reception

As mentioned previously, the influence of short-term fading is mitigated through the usage of diversity techniques typically at the single base station (microdiversity). However, the usage of such microdiversity approaches alone will not be sufficient to mitigate the overall wireless channel degradation, caused by simultaneous influence of short-term and long-term fading (shadowing). The technique used to alleviate the effects of shadowing, which can put a heavy limit on system performances along with the effects of short-time fading, is called *macrodiversity*. Here, multiple signals are received at geographically distanced base stations (BS), ensuring that different long-term fading is experienced by these signals [28]. The simultaneous usage of multiple BSs and the processing of signals from these BSs will provide the framework for both macro- and microdiversity techniques to improve the performance in shadowed fading channels. The most commonly used macrodiversity combining scheme is based on the simple selection of the BS with the largest mean signal power value.

For example, the CDF at the output of the macrodiversity combiner, consisting of two BSs, can be determined according to the relation [29]

$$
F_z(z) = \int_0^\infty d\Omega_1 \int_0^{\Omega_1} d\Omega_2 F_{z_1/\Omega_1}\left(\frac{z}{\Omega_1}\right) f_{\Omega_1\Omega_2}(\Omega_1\Omega_2)
$$

$$
+ \int_0^\infty d\Omega_2 \int_0^{\Omega_2} d\Omega_1 F_{z_2/\Omega_2}\left(\frac{z}{\Omega_2}\right) f_{\Omega_1\Omega_2}(\Omega_1\Omega_2), \qquad (4.35)
$$

with $F(z_i/\Omega_i)$ being the CDF of the signal envelope at the output of ith $(i = 1,2)$ microcombiner (BS), conditioned over mean (average) signal power value Ω_i $(i = 1,2)$, while $f_{\Omega_1\Omega_2}(\Omega_1\Omega_2)$ stands for the JPDF of mean signal powers at the macrocombiner inputs and models the shadowing process to which the system has been exposed. In each time instance, the selection is based on the largest mean signal power value: if $\Omega_1 > \Omega_2$ then first BS is chosen, while in the case when $\Omega_2 > \Omega_1$ the second BS is selected.

In a similar manner, other performance measures at the macrodiversity output can be determined, and further analysis of macrodiversity system performances is provided in Chapter 7.

References

1. Simon, M. K. and Alouini, M. S. (2005). *Digital Communications over Fading Channels*, 2nd edn. Wiley, New York.
2. Stavroulakis, P. (2003). *Interference Analysis and Reduction for Wireless Systems*. Artech House, London, U.K.
3. Alouini, M. and Goldsmith A. (1999). Area spectral efficiency of cellular mobile radio systems. *IEEE Transactions on Vehicular Technology*, 48(4), 1047–1065.
4. Tellambura, C., Annamalai, A., and Bhargava V. (2001). Unified analysis of switched diversity systems in independent and correlated fading channels. *IEEE Transactions on Communications*, 49(11), 1955–1965.
5. Tellambura, C. and Bhargava, V. (1994). Unified error analysis of DQPSK in fading channels. *Electronic Letters*, 30(25), 2110–2111.
6. Pawula, R. (1998). A new formula for MDPSK symbol error probability. *IEEE Communication Letters*, 2(10), 271–272.

7. Dong, X., Beaulieu, N. C., and Wittke, P. H. (1998). Two dimensional signal constellations for fading channels. *Proceedings of the IEEE GLOBECOM, Communication Theory Mini Conference*, San Francisco, pp. 22–27.

8. Gradshteyn, I. and Ryzhik, I. (1980). *Tables of Integrals, Series, and Products*. Academic Press, New York.

9. Alouini, M. S. and Simon, M. (2000). An MGF-based performance analysis of generalized selection combining over Rayleigh fading channels. *IEEE Transactions on Communications*, 48(3), 401–415.

10. Alouini, M. S. and Goldsmith, A. J. (1999). A unified approach for calculating the error rates of linearly modulated signals over generalized fading channels. *IEEE Transactions on Communications*, 47(9), 1324–1334.

11. Dong, X., Beaulieu, N. C., and Wittke, P. H. (1999). Signaling constellations for fading channels. *IEEE Transactions on Communications*, 47(5), 703–714.

12. Charash, U. (1979). Reception through Nakagami fading multipath channels with random delays. *IEEE Transactions on Communications*, 27(4), 657–670.

13. Alouini, M. S. and Simon, M. (2001). Dual diversity over log-normal fading channels. *Proceedings of the IEEE International Conference on Communications*, Helsinki, Finland, Vol. 4, pp. 1089–1093.

14. Lee, W. C. Y. (1967). Statistical analysis of the level crossings and duration of fades of the signal from an energy density mobile radio antenna. *Bell System Technical Journal*, 46(2), 412–448.

15. Mitic, A. and Jakovljevic M. (2007). Second-order statistics in Weibull-lognormal fading channels. *Proceedings of Telecommunications in Modern Satellite, Cable and Broadcasting Services Conference*, Nis, Serbia, pp. 26–28.

16. Struber, G. L. (1996). *Principles of Mobile Communications*. Kluwer Academic Publishers, Boston, MA.

17. Rappaport, T. S. (1996). *Wireless Communications: Principles and Practice*. PTR Prentice-Hall, Upper Saddle River, NJ.

18. Brennan, D. (1959). Linear diversity combining techniques. *Proceedings of Institute of Radio Engineers*, 47, 1075–1102.

19. Fraidenraich, G., Filho, J., and Yacoub, M. D. (2005). Second-order statistics of maximal-ratio and equal-gain combining in Hoyt fading. *IEEE Communications Letters*, 9(1), 19–21.

20. Neasmith, E. A. and Beaulieu N. C. (1998). New results in selection diversity. *IEEE Transactions on Communications*, 46(5), 695–704.

21. Okui, S. (2000). Effects of SIR selection diversity with two correlated branches in the m-fading channel. *IEEE Transactions on Communications*, 48(10), 1631–1633.

22. Austin, D. and Stuber, L. (1995). In-service signal quality estimation for TDMA cellular systems. *Proceedings of the IEEE International Symposium on Personal, Indoor and Mobile Radio Communications*, Toronto, Canada, pp. 836–840.

23. Brandao, A., Lopez, L., and McLernon, C. (1994). Co-channel interference estimation for M-ary PSK modulated signals. *Wireless Personal Communications*, 1(1), 23–32.

24. Helstrom, C. W. (1991). *Probability and Stochastic Processes for Engineers*, 2nd edn. MacMillan, London, U.K.
25. Abu-Dayya A. A. and Beaulieu, N. C. (1994). Analysis of switched diversity systems on generalized-fading channels. *IEEE Transactions on Communications*, 42(11), 2959–2966.
26. Ko, Y. C., Alouni, M. S., and Simon, M. K. (2000). Analysis and optimization of switched diversity systems. *IEEE Transactions on Communications*, 49(5), 1813–1831.
27. Bithas, P. S., Mathiopoulos, P. T., and. Karagiannidis, G. K. (2006). Switched diversity receivers over correlated Weibull fading channels. *Proceedings of the International Workshop on Satellite and Space Communications*, Madrid, Spain, pp. 143–147, September 2006.
28. Bernhardt, R. C. (1987). Macroscopic diversity in frequency reuse radio systems. *IEEE Journal on Selected Areas in Communications*, 5(5), 862–870.
29. Stefanovic, D., Panic, S., and Spalevic, P. (2011). Second order statistics of SC macrodiversity system operating over gamma shadowed Nakagami-m fading channels. *International Journal of Electronics and Communications (AEUE)*, 65(5), 413–418.

5

SINGLE-CHANNEL RECEIVER OVER FADING CHANNELS IN THE PRESENCE OF CCI

In order to point out both the necessity and the validity of space diversity techniques usage, from the point of view of multipath fading and cochannel interference (CCI) influence mitigation, let us consider in this chapter performances of single-channel receiver over multipath fading channel in the presence of CCI. Performance evaluations that follow will consider multichannel reception, along with the presentation of appropriate expressions for evaluating the standard performance measures of these realistic systems over the CCI, such as outage probability (OP) and symbol error probability (SEP) of the various modulation/detection schemes defined in the literature. Most of the results in this chapter are originally reported by the contributing authors in the literature [1–5], and the reader who is interested in more details of the derivations is referred to the references.

5.1 Performance Analysis of Reception over α-μ Fading Channels in the Presence of CCI

Let us consider the case with transmitted signal exposed to the influence of multipath fading, modeled by α-μ distribution, and to the influence of CCI. We will provide the probability density function (PDF) and cumulative distribution function (CDF) of the received signal-to-interference ratio (SIR) in closed form, in this section.

The desired signal, received by the single antenna, can be written as [6,7]

$$D(t) = R(t)e^{j\phi(t)}e^{j\left[2\pi f_c t + \Phi(t)\right]}, \tag{5.1}$$

where
  f_c denotes the carrier frequency
  $\Phi(t)$ is the desired information signal

$\phi(t)$ is the random phase uniformly distributed in $[0.2\pi]$

$R(t)$ is the α-μ modeled random process, given by [8]

$$f_R(R) = \frac{\alpha \mu_d^{\mu_d} R^{\alpha\mu_d-1}}{\Gamma(\mu_d)\hat{R}^{\alpha\mu_d}} \exp\left(-\mu_d \frac{R^\alpha}{\hat{R}^\alpha}\right), \quad R \geq 0, \tag{5.2}$$

where

$\Gamma(x)$ stands for the gamma function [9, Eq. (8.310⁷.1)]

as reported previously, α is related to the nonlinearity of the environment

μ_d parameter is associated with the number of multipath clusters through which the desired signal propagates

$\hat{R} = \sqrt[\alpha]{E(R^\alpha)}$ being the α-root signal's mean value

Similarly, the resultant interfering signal received by the single antenna is provided by

$$C(t) = r(t)e^{j\theta(t)}e^{j[2\pi f_c t+\psi(t)]}, \tag{5.3}$$

where $r(t)$ is also α-μ distributed random amplitude process of CCI, given by

$$f_r(r) = \frac{\alpha \mu_c^{\mu_c} r^{\alpha\mu_c-1}}{\Gamma(\mu_c)\hat{r}^{\alpha\mu_d}} \exp\left(-\mu_c \frac{r^\alpha}{\hat{r}^\alpha}\right), \quad r \geq 0, \tag{5.4}$$

where

$\theta_i(t)$ is the random phase

$\psi(t)$ is the information signal

Let us define the instantaneous value of the SIR as $\lambda = R^2/r^2$. The PDF of this ratio can be determined, according to [7], as

$$f_\lambda(\lambda) = \frac{1}{4\sqrt{\lambda}} \int_0^\infty f_R\left(r\sqrt{\lambda}\right) f_r(r) r\, dr. \tag{5.5}$$

After substituting (3.2) and (3.4) into (3.5), the PDF of the single-channel receiver SIR can be presented as

$$f_\lambda(\lambda) = \frac{\alpha \mu_d^{\mu_d} \mu_c^{\mu_c} \lambda^{(\alpha\mu_d/2)-1}}{2\left(\mu_c S^{(\alpha/2)} + \mu_d \lambda^{(\alpha/2)}\right)^{\mu_d+\mu_c}} \frac{\Gamma(\mu_d+\mu_c)}{\Gamma(\mu_d)\Gamma(\mu_c)}, \tag{5.6}$$

where the average value of SIR is denoted as $S = \hat{R}^2/\hat{r}^2$.

Similarly, the CDF of instantaneous received SIR can be obtained as

$$F_\lambda(\lambda) = \int_0^\lambda f_t(t)\,dt;$$

$$F_\lambda(\lambda) = \frac{\Gamma(\mu_d + \mu_c)}{\Gamma(\mu_d)\Gamma(\mu_c)}\frac{\left(\mu_d\lambda^{\alpha/2}/(\mu_c S^{\alpha/2} + \mu_d\lambda^{\alpha/2})\right)^{\mu_d}}{\mu_d}$$

$$\times\; _2F_1\left(\mu_d, 1-\mu_c, 1+\mu_d, \frac{\mu_d\lambda^{\alpha/2}}{\mu_c S^{\alpha/2} + \mu_d\lambda^{\alpha/2}}\right), \qquad (5.7)$$

with $_2F_1(u_1,u_2;u_3;x)$ being the Gaussian hypergeometric function [9, Eq. (9.14)]. Capitalizing on (5.7), according to the definition provided in Chapter 4, the OP for this transmission scenario has been efficiently evaluated and presented in Figure 5.1, as the function of normalized average SIR value S/γ. It is obvious that OP values are extremely high in the area of strong influence of CCI, that is, $S_1/\gamma < 3$ dB. In addition, it is evident that both the α parameter (fading severity decrease) and the μ parameter (number of multipath propagation components of same phase delay) growth leads to OP decrease, that is, to the system performance improvement.

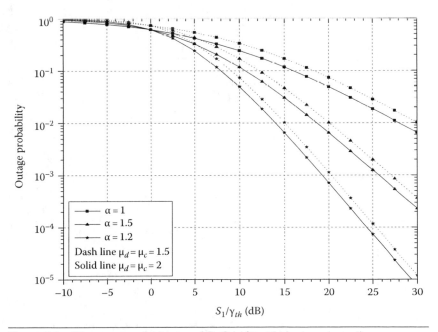

Figure 5.1 OP versus normalized average SIR, S_1/γ, for a single-channel reception.

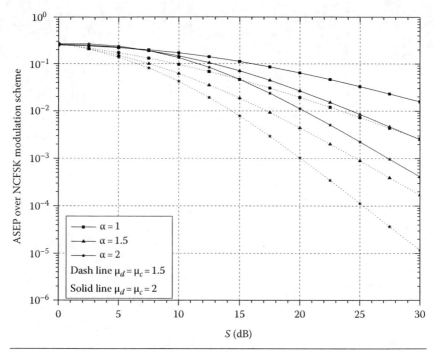

Figure 5.2   ASEP over NCFSK modulation scheme.

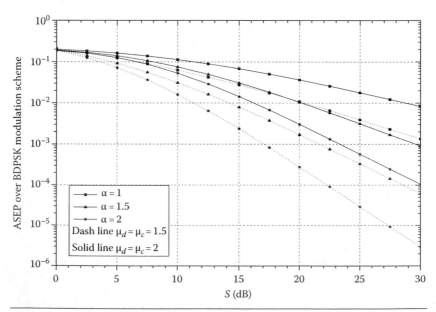

Figure 5.3   ASEP over BDPSK modulation scheme.

In a similar manner, another important performance measure average symbol error probability (ASEP) was considered, capitalizing on (5.6) and (4.7). ASEP values for a single-channel receiving structure operating over BDPSK and NCFSK modulation formats in wireless environment exposed to the influence of multipath α-μ fading and CCI are efficiently evaluated and presented at Figures 5.2 and 5.3. ASEP values are presented in the function of average received SIR. Beyond the conclusions derived about the influence of α and μ parameter growth on the performance improvement, we can also observe that lower ASEP values are obtained when the BDPSK modulation format is applied, compared to the case when the transmission with NCFSK modulation format is used.

5.2 Performance Analysis of the Reception over κ-μ Fading Channels in the Presence of CCI

Let us consider the case when transmission is carried out over a κ-μ fading environment in the presence of CCI. Closed-form expressions for received SIR PDF and CDF will be derived, as well as further useful closed-form expressions for ASEP over some coherent and some noncoherent modulation systems.

The desired signal envelope in a κ-μ fading channel follows the PDF given by [10]

$$f_R(R) = \frac{2\mu_d \left(1+k_d\right)^{(\mu_d+1)/2} R^{\mu_d}}{k_d^{(\mu_d-1)/2} e^{\mu_d k_d} \Omega_d^{(\mu_d+1)/2}}$$

$$\times \exp\left(-\frac{\mu_d \left(1+k_d\right) R^2}{\Omega_d}\right) I_{\mu_d-1}\left[2\mu_d \sqrt{\frac{k_d \left(1+k_d\right) R^2}{\Omega_d}}\right], \quad (5.8)$$

where

$\Omega_d = E\left[R^2\right]$ is the desired signal average power

$I_0\left(x\right)$ is the zeroth-order-modified Bessel function of the first kind [9, Eq. (8.406)]

It has already been explained that parameter k_d is the ratio between the total power of dominant components and the total power of scattered waves [11]. Similarly, the CCI signal envelope in a κ-μ fading channel follows

$$f_r(r) = \frac{2\mu_c \left(1+k_c\right)^{(\mu_c+1)/2} r^{\mu_c}}{k_c^{(\mu_c-1)/2} e^{\mu_c k_c} \Omega_c^{(\mu_c+1)/2}}$$

$$\times \exp\left(-\frac{\mu_c\left(1+k_c\right)r^2}{\Omega_c}\right) I_{\mu_c-1}\left[2\mu_c\sqrt{\frac{k_c\left(1+k_c\right)r^2}{\Omega_c}}\right], \qquad (5.9)$$

with parameters defined in a similar manner. Since an interference-limited system is considered, the effect of noise could be ignored. In this case, it can be shown that the instantaneous SIR, $\lambda = R^2/r^2$, has the PDF

$$f_\lambda(t) = \frac{1}{2\sqrt{t}} \int_0^\infty f_R\left(r\sqrt{t}\right) f_r(r) r\, dr;$$

$$f_\lambda(\lambda) = \sum_{p=0}^\infty \sum_{q=0}^\infty \left(\frac{\lambda^{p+\mu_d-1} S^{\mu_c+q} \mu_d^{2p+\mu_d} \mu_c^{2q+\mu_c} k_d^p k_c^q}{e^{k_d\mu_d+k_c\mu_c}\Gamma\left(p+\mu_d\right)p!\Gamma\left(q+\mu_c\right)q!}\right.$$

$$\left. \times \frac{(1+k_d)^{\mu_d+p}(1+k_c)^{\mu_c+q}\Gamma(p+\mu_d+q+\mu_c)}{\left(\lambda(1+k_d)\mu_d + S(1+k_c)\mu_c\right)^{p+\mu_d+q+\mu_c}}\right), \qquad (5.10)$$

with $S = \Omega_d/\Omega_c$ being the average signal to average SIR, a measure useful in determining the cochannel reduction factor in systems with frequency reuse.

Now, the CDF of λ can be obtained as

$$F_\lambda(t) = \sum_{p=0}^\infty \sum_{q=0}^\infty \left[\frac{\mu_d^p \mu_c^q k_d^p k_c^q \Gamma(p+\mu_d+q+\mu_c)}{e^{k_d\mu_d+k_c\mu_c}\Gamma\left(p+\mu_d\right)p!\Gamma\left(q+\mu_c\right)q!}\right.$$

$$\times \frac{\left(\lambda(1+k_d)\mu_d \big/ \left(\lambda(1+k_d)\mu_d + S(1+k_c)\mu_c\right)\right)^{p+\mu_d}}{p+\mu_d}$$

$$\times {_2F_1}\left(p+\mu_d, 1-q-\mu_c;\; p+\mu_d+1;\right.$$

$$\left.\left. \times \frac{\lambda(1+k_d)\mu_d}{\left(\lambda(1+k_d)\mu_d + S(1+k_c)\mu_c\right)}\right)\right]. \qquad (5.11)$$

In an interference-limited environment, ASEP can be derived by averaging the conditional error probability P_e over the PDF of SIR:

$$\bar{P}_e = \int_0^\lambda P_e(\lambda) f_\lambda(\lambda) d\lambda; \quad P_e(\lambda) = \frac{1}{2} \exp(-g\lambda);$$

$$\bar{P}_e = \sum_{k=0}^\infty \sum_{l=0}^\infty \left[\frac{\mu_d^p \mu_c^q k_d^p k_c^q \Gamma(p + \mu_d + q + \mu_c)}{2 e^{k_d \mu_d + k_c \mu_c} \, p! \Gamma(q + \mu_c) q!} \right.$$

$$\left. \times \psi \left(p + \mu_d; -q - \mu_c; \frac{\alpha S(1 + k_c)\mu_c}{\lambda(1 + k_d)\mu_d} \right) \right], \tag{5.12}$$

where in noncoherent systems, for $\alpha = 1$ binary PSK, and for $\alpha = 1/2$ binary FSK format is modeled. $\psi(a, b, x)$ is the confluent hypergeometric function of the second kind [9, Eq. (3.383.5[7])].

For a coherent system, the conditional probability of error may be expressed in terms of the confluent hypergeometric function [9, Eq. (9.14)] as

$$P_e = \frac{1}{2} \operatorname{erfc}(\sqrt{\alpha\lambda}) = \frac{1}{2} \left(1 - 2\sqrt{\frac{\alpha\lambda}{\pi}} \, {}_1F_1\left(\frac{1}{2}; \frac{3}{2}; -\alpha\lambda \right) \right);$$

$$\bar{P}_e = \frac{1}{2} - \sqrt{\frac{\alpha}{\pi}} \sum_{p=0}^\infty \sum_{q=0}^\infty \sum_{k=0}^\infty \frac{(-\alpha)^k (1/2)_k \Gamma(2k + 2l + 2)}{(3/2)_k \, p! \Gamma(k+1) k! \Gamma(l+1) l!}$$

$$\times \frac{\mu_d^{p+k} \mu_c^{q+k} (1 + k_d)^k (1 + k_c)^k k_d^p k_c^q S^k \Gamma(p + \mu_d + q + \mu_c)}{e^{k_d \mu_d + k_c \mu_c} \, p! \Gamma(q + \mu_c) \Gamma(p + \mu_d) q!}$$

$$\times \frac{\left(\lambda(1 + k_d)\mu_d / \left(\lambda(1 + k_d)\mu_d + S(1 + k_c)\mu_c \right) \right)^{p + \mu_d}}{p + \mu_d}$$

$$\times {}_2F_1\left(p + \mu_d, 1 - q - \mu_c; p + \mu_d + 1; \frac{\lambda(1 + k_d)\mu_d}{\left(\lambda(1 + k_d)\mu_d + S(1 + k_c)\mu_c \right)} \right), \tag{5.13}$$

where $(a)_p$ is the Pochhammer symbol [9] and α was previously defined for required modulation techniques as $\alpha = 1$ for CPSK, and $\alpha = 1/2$ for CFSK.

ASEP versus average SIR for binary DPSK and NCFSK in a κ-μ fading environment for various values of system parameters are presented in Figures 5.4 and 5.5, respectively. A balanced input SIR $(S_1 = S_2 = S)$ system was observed. Here we can observe obvious upgrading of system performances when higher values of fading severity parameters κ and μ better performances are present.

5.3 Performance Analysis of the Reception over Hoyt Fading Channels in the Presence of CCI

Closed-form expressions for standard performance measures of the receiver operating over Hoyt (Nakagami-q) fading in the presence of CCI will be provided in this section. The desired signal envelope in the Hoyt fading channel follows the PDF given by [12].

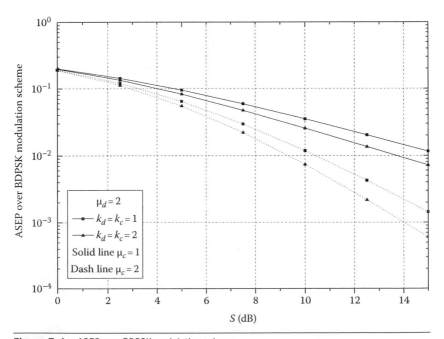

Figure 5.4 ASEP over BDPSK modulation scheme.

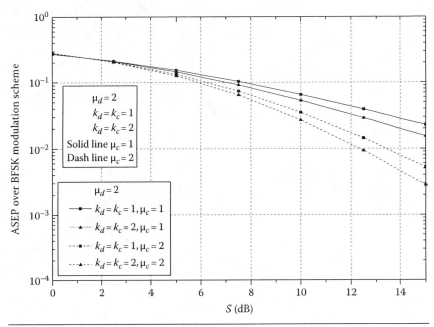

Figure 5.5   ASEP over BFSK modulation scheme.

$$f_R(R) = \frac{(1+q_d^2)R}{q_d\Omega_d} \exp\left(-\frac{(1+q_d^2)R}{4q_d^2\Omega_d}\right) I_0\left(\frac{(1-q_d^4)R}{4q_d^2\Omega_d}\right), \qquad (5.14)$$

where

$\Omega_d = E[R^2]$ is the desired signal average power

$I_0(x)$ is the zeroth-order-modified Besel function of the first kind

q_d, $0 \le q_d \le 1$, is the desired signal Hoyt fading parameter

As mentioned previously, the Hoyt distribution spans the range from one-sided Gaussian fading ($q_d = 0$) to Rayleigh fading ($q_d = 1$).

Similarly, the CCI envelope in the Nakagami-q fading channel follows

$$f_r(r) = \frac{(1+q_c^2)r}{q_c\Omega_c} \exp\left(-\frac{(1+q_c^2)r}{4q_c^2\Omega_c}\right) I_0\left(\frac{(1-q_c^4)r}{4q_c^2\Omega_c}\right), \qquad (5.15)$$

with parameters defined in a similar manner.

In an interference-limited system, the effect of noise may be ignored, and in this case, it can be shown that the instantaneous SIR, $\lambda = R^2/r^2$, would have the PDF in the following form:

$$f_\lambda(t) = \frac{1}{2\sqrt{t}} \int_0^\infty f_R\left(r\sqrt{t}\right) f_r(r) r\, dr;$$

$$f_\lambda(t) = \sum_{k=0}^\infty \sum_{l=0}^\infty \left(\frac{S^{2l+1}\left(1-q_d^4\right)^{2k}\left(1-q_c^4\right)^{2l}\left(1+q_d^2\right)\left(1+q_c^2\right)}{2^{2k+2l+2}\Gamma(k+1)k!\Gamma(l+1)l!} \right.$$

$$\left. \times \frac{q_d^{4l+3} q_c^{4k+3} \lambda^{2k}}{\left(\lambda\left(1+q_d^2\right)^2 q_c^2 + S\left(1+q_c^2\right)^2 q_d^2\right)^{2k+2l+2}} \right),$$ (5.16)

where the ratio $S = \Omega_d/\Omega_c$ stands for the average SIR, which is useful in determining the cochannel reduction factor in systems with frequency reuse.

The CDF of λ can be obtained as

$$F_\lambda(\lambda) = \int_0^\lambda f_t(t)\, dt$$

$$= \sum_{k=0}^\infty \sum_{l=0}^\infty \left(\frac{S^{2l+1}\left(1-q_d^4\right)^{2k}\left(1-q_c^4\right)^{2l} q_d q_c}{2^{2k+2l+2}\Gamma(k+1)k!\Gamma(l+1)l!\left(1+q_d^2\right)^{4k+1}\left(1+q_c^2\right)^{4l+1}} \right.$$

$$\times \frac{\left(\lambda\left(1+q_d^2\right)^2 q_c^2 / \lambda\left(1+q_d^2\right)^2 q_c^2 + S\left(1+q_c^2\right)^2 q_d^2\right)^{2k+2l+2}}{2k+1}$$

$$\left. \times {}_2F_1\left(2k+1,-2l; 2k+1; \frac{\lambda\left(1+q_d^2\right)^2 q_c^2}{\lambda\left(1+q_d^2\right)^2 q_c^2 + S\left(1+q_c^2\right)^2 q_d^2}\right) \right).$$

(5.17)

For a noncoherent system, in an interference-limited environment, ASEP can be derived by averaging the conditional error probability P_e over the PDF of SIR:

$$\bar{P}_e = \int_0^\lambda P_e(\lambda) f_\lambda(\lambda) d\lambda; \quad P_e(\lambda) = \frac{1}{2}\exp(-g\lambda);$$

$$\bar{P}_e = \sum_{k=0}^\infty \sum_{l=0}^\infty \left(\frac{\left(1-q_d^4\right)^{2k}\left(1-q_c^4\right)^{2l} q_d q_c \Gamma(2k+2l+2)\Gamma(2k+1)}{2^{2k+2l-1}\Gamma(k+1)k!\Gamma(l+1)l!\left(1+q_d^2\right)^{4k+1}\left(1+q_c^2\right)^{4l+1}} \right.$$

$$\left. \times \psi\left(2k+1; -2l; \frac{\alpha S\left(1+q_c^2\right)^2 q_d^2}{\lambda\left(1+q_d^2\right)^2 q_c^2} \right) \right), \tag{5.18}$$

with $\alpha=1$ for binary PSK, and with $\alpha=1/2$ for binary FSK. For a coherent system, the conditional probability of error may be expressed in terms of the confluent hypergeometric function [10]:

$$P_e = \frac{1}{2} \mathrm{erfc}\left(\sqrt{\alpha\lambda}\right) = \frac{1}{2}\left(1 - 2\sqrt{\frac{\alpha\lambda}{\pi}} \;_1F_1\left(\frac{1}{2};\frac{3}{2};-\alpha\lambda\right) \right);$$

$$\bar{P}_e = \frac{1}{2} - \sqrt{\frac{\alpha\lambda}{\pi}} \sum_{k=0}^\infty \sum_{l=0}^\infty$$

$$\times \sum_{p=0}^\infty \left(\frac{S^{p+(1/2)}\left(1-q_d^4\right)^{2k}\left(1-q_c^4\right)^{2l} q_d^{2p+2}\,\Gamma(2k+p+(3/2))}{2^{2k+2l-2}\left(1+q_d^2\right)^{4k+1}\left(1+q_c^2\right)^{4l+1} q_c^{2p}} \right.$$

$$\times \frac{\alpha^p (1/2)_p \Gamma(2k+2l+2)}{(3/2)_p\, p!\Gamma(k+1)k!\Gamma(l+1)l!}$$

$$\left. \times \psi\left(2k+p+\frac{3}{2};\, p+\frac{1}{2}-2l;\, \frac{\alpha S\left(1+q_c^2\right)^2 q_d^2}{\lambda\left(1+q_d^2\right)^2 q_c^2} \right) \right), \tag{5.19}$$

where α was previously defined for required modulation techniques. Detailed derivations are presented in [4].

5.4 Performance Analysis of the Reception over η-μ Fading Channels in the Presence of CCI

The case of small-scale signal variation in general non-line-of-sight condition, where in-phase and quadrature components of the fading signal within each cluster are assumed to be independent mutually, and having different powers, will be observed in this section. The CCI influence will also be observed. The desired signal η-μ distributed random amplitude process can be modeled [13]

$$f_R(R) = \frac{4\sqrt{\pi}\mu_d^{\mu_d+1/2}\,h_d^{\mu_d}\,R^{2\mu_d}}{\Gamma(\mu_d)H_d^{\mu_d-1/2}\,\Omega_d^{\mu_d+1/2}}$$

$$\times \exp\left(-\frac{2\mu_d h_d R^2}{\Omega_d}\right) I_{\mu_d-1/2}\left[\frac{2\mu_d H_d R^2}{\Omega_d}\right]; \qquad (5.20)$$

where

$\Omega_d = E[R^2]$ is the desired signal average power

H_d and h_d are parameters defined in the function of η_d as [13]

$$H_d = \frac{\eta_d^{-1} - \eta_d}{4}; \quad h_d = \frac{2 + \eta_d^{-1} + \eta_d}{4}. \qquad (5.21)$$

Roughly speaking, the parameter μ_d is related to the number of multipath clusters in the environment, whereas the parameter η_d is related to the scattered wave power ratio between the in-phase and quadrature components of each cluster of multipath [14]. The generality of this model is evident, because, as special cases, it includes other fading models, such as Nakagami-q (Hoyt), one-sided Gaussian, Rayleigh, and Nakagami-m. For $\mu_d = 0.5$, it reduces to the Nakagami-q model, with parameter q corresponding to $\eta_d^{1/2}$. The Nakagami-m model is approximated when $\eta_d \to 0$ and $\eta_d \to \infty$, with m corresponding to μ_d.

In the same manner, the resultant interfering signal received by the antenna is also η-μ distributed random amplitude process and can be presented as

$$f_r(r) = \frac{4\sqrt{\pi}\mu_c^{\mu_c+1/2}\,h_c^{\mu_c}\,r^{2\mu_c}}{\Gamma(\mu_c)H_c^{\mu_c-1/2}\Omega_c^{\mu_c+1/2}}\,\exp\left(-\frac{2\mu_c h_c R^2}{\Omega_c}\right) I_{\mu_c-1/2}\left[\frac{2\mu_c H_c r^2}{\Omega_c}\right],$$

$$(5.22)$$

where $\Omega_c = E[R^2]$ is the interfering signal average power.

Now, with respect to (5.20) and (5.22), the instantaneous SIR, $\lambda = R^2/r^2$, has the PDF

$$f_\lambda(t) = \frac{1}{2\sqrt{t}} \int_0^\infty f_R\left(r\sqrt{t}\right) f_r(r) r\, dr;$$

$$f_\lambda(\lambda) = \sum_{k=0}^\infty \sum_{l=0}^\infty \frac{\pi S^{2k+2\mu_c} \mu_c^{2k+2\mu_c} \mu_d^{2l+2\mu_d} b_d^{\mu_d} b_c^{\mu_c} H_d^{2l} H_c^{2k}}{2^{2k++2l+2\mu_c+2\mu_d-2} \Gamma(\mu_d)\Gamma(\mu_c)\Gamma(\mu_d+k+1/2)}$$

$$\times \frac{\Gamma(2k+2l+2\mu_c+2\mu_d)\lambda^{2l+2\mu_d-1}}{\Gamma(\mu_c+l+1/2)\,k!\,l!\,(\lambda\mu_d b_d + S\mu_c b_c)^{2k+2l+2\mu_c+2\mu_d}},$$

(5.23)

where $S = \Omega_d/\Omega_c$ denotes the average SIR value at the receiver. Following (5.23), the CDF of λ can be obtained as

$$F_\lambda(t) = \sum_{p=0}^\infty \sum_{q=0}^\infty \frac{\pi H_d^{2l} H_c^{2k} \Gamma(2k+2l+2\mu_c+2\mu_d)}{2^{2k++2l+2\mu_c+2\mu_d-2} b_d^{2l+\mu_d} b_c^{2k+\mu_c} \Gamma(\mu_d)\Gamma(\mu_c)}$$

$$\times \frac{B_z(2l+2\mu_d, 2k+2\mu_c)}{\Gamma(\mu_d+k+1/2)\Gamma(\mu_c+l+1/2)\,k!\,l!};$$

$$z = \frac{\lambda\mu_d b_d}{\lambda\mu_d b_d + S\mu_c b_c},$$

(5.24)

with $B(z,a,b)$ being the incomplete beta function [9, Eq. (8.391)].

For a noncoherent system, in an interference-limited environment, ASEP can be expressed as

$$\bar{P}_e = \sum_{k=0}^\infty \sum_{l=0}^\infty \frac{\pi H_d^{2l} H_c^{2k} \Gamma(2k+2l+2\mu_c+2\mu_d)\Gamma(2l+2\mu_d)}{2^{2k++2l+2\mu_c+2\mu_d-1} b_d^{2l+\mu_d} b_c^{2k+\mu_c} \Gamma(\mu_d)\Gamma(\mu_c)}$$

$$\times \frac{\psi\left(2l+2\mu_d; 1-2k+2\mu_c; \alpha S\mu_c b_c/\lambda\mu_d b_d\right)}{\Gamma(\mu_d+k+1/2)\Gamma(\mu_c+l+1/2)\,k!\,l!},$$

(5.25)

with $\alpha = 1$ for binary PSK, and with $\alpha = 1/2$ for binary FSK.

For a coherent system, ASEP can be expressed as

$$
\bar{P}_e = \sum_{k=0}^{\infty} \sum_{l=0}^{\infty} \frac{(S\mu_c)^{2k-2l+2\mu_c-2\mu_d} \mu_d^{2l-2k-2\mu_c+2\mu_d} \pi H_d^{2l} H_c^{2k}}{2^{2k++2l+2\mu_c+2\mu_d-1} b_d^{2l+\mu_d-\mu_c} b_c^{2k+\mu_c-\mu_d} \Gamma(\mu_d)\Gamma(\mu_c)}
$$

$$
\times \frac{\Gamma(2k+2\mu_c)\Gamma(2l+2\mu_d)}{\Gamma(\mu_d+k+1/2)\Gamma(\mu_c+l+1/2)k!l!}
$$

$$
- \sum_{k=0}^{\infty} \sum_{l=0}^{\infty} \sqrt{\pi}(-1)^p g^{2p+1} \frac{(S\mu_c)^{2k-2l+2\mu_c-2\mu_d-2p-1} \mu_d^{2l-2k-2\mu_c+2\mu_d-2p-1}}{2^{2k++2l+2\mu_c+2\mu_d-2} b_d^{2l+\mu_d-\mu_c+2p+1} b_c^{2k+\mu_c-\mu_d}}
$$

$$
\times \frac{\pi H_d^{2l} H_c^{2k} \Gamma(2k+2\mu_c)\Gamma(2l+2\mu_d)}{\Gamma(\mu_d)\Gamma(\mu_c)\Gamma(\mu_d+k+1/2)\Gamma(\mu_c+l+1/2)k!l!p!}, \tag{5.26}
$$

where α was previously defined for required modulation techniques. Detailed derivations are presented in [3].

5.5 Performance Analysis of the Reception over α-η-μ Fading Channels in the Presence of CCI

The analytical framework for performance analysis of a wireless communication system subjected to CCI over α-η-μ fading channels will be presented in this section. SIR-based analysis will be provided, and closed-form expressions will be derived for received SIR PDF and CDF. From this statistics, OP values will be obtained in the function of system parameters.

Recently, the nonlinear properties of propagation medium have been considered extensively. Various short-time fading distributions, like Nakagami-m, Rician, and Rayleigh, assume a resultant homogenous diffuse scattering field, from randomly distributed scatters. However, surfaces are often spatially correlated and they characterize a nonlinear environment. Exploring the fact that the resulting envelope would be a nonlinear function of the sum of multipath components, a novel, general α-η-μ distribution for a short-time fading model was recently presented. PDF is presented in the form of three parameters α, μ and η, which are related to the nonlinearity of the environment, the number of multipath clusters in the environment, and the scattered wave power ratio between the in-phase and quadrature components of each cluster of multipath, respectively [13].

Due to a general fading distribution, the α-η-μ model includes other short-time fading distributions, such as Rayleigh, Nakagami-q (Hoyt), Nakagami-m, η-μ, Weibull, and one-sided Gaussian distribution, as special cases. By setting parameter α to value $\alpha = 2$, it reduces to η-μ distribution. Further, from the η-μ fading distribution the Nakagami-m model could be obtained in two cases: first for $\eta \to 1$, with parameter m being expressed as $m = \mu/2$, and second for $\eta \to 0$, with parameter m being expressed as $m = \mu$. It is well known that η-μ distribution reduces to the Hoyt distribution, for the case when $\mu = 1$, with parameter b defined as $b = (1 - \eta)/(1 + \eta)$. By equating the in-phase and quadrature components variances, namely by setting $\eta = 1$, the Rayleigh distribution is derived from Hoyt. Also, the Weibull distribution could be obtained as special case of α-η-μ model by setting corresponding values to the parameters $\mu = 1$ and $\eta = 1$. Obviously, the major contribution of this analysis is the previously mentioned generality.

The obtained mathematical form will allow simple performance analysis of wireless communication systems, operating in composite fading environments. SIR- based performance analysis is a very effective performance criterion, since SIR can be measured in real time both in the base and in the mobile station(s). An interference-limited system will be discussed, so the effect of noise will be ignored.

The desired information signal with a α-η-μ distributed random amplitude process can be presented by [13]

$$f_R(R) = \frac{\alpha(\eta_d - 1)^{(1/2)-\mu_d} (\eta_d + 1)^{(1/2)+\mu_d} \sqrt{\pi}\mu_d^{(1/2)+\mu_d} R^{\alpha((1/2)+\mu_d)-1}}{\sqrt{\eta_d}\,\Gamma(\mu_d)\Omega_d^{(1/2)+\mu_d}}$$
$$\times \exp\left(-\frac{(1+\eta_d)^2}{2\eta_d}\frac{R^\alpha}{\Omega_d}\right)I_{\mu_d-1/2}\left(\frac{(\eta_d^2-1)}{2\eta_d}\frac{R^\alpha}{\Omega_d}\right), \qquad (5.27)$$

with $\Omega_d = E[R^2]$ denoting the desired signal average power.

In a similar manner, the interfering signal can be presented as

$$f_r(r) = \frac{\alpha(\eta_c - 1)^{(1/2)-\mu_c} (\eta_c + 1)^{(1/2)+\mu_c} \sqrt{\pi}\mu_c^{(1/2)+\mu_c} r^{\alpha((1/2)+\mu_c)-1}}{\sqrt{\eta_d}\,\Gamma(\mu_d)\Omega_d^{(1/2)+\mu_d}}$$
$$\times \exp\left(-\frac{(1+\eta_c)^2}{2\eta_c}\frac{r^\alpha}{\Omega_c}\right)I_{\mu_c-1/2}\left(\frac{(\eta_c^2-1)}{2\eta_c}\frac{r^\alpha}{\Omega_c}\right), \qquad (5.28)$$

with $\Omega_c = E[r^2]$ denoting the CCI signal average power, while parameters α, μ_c and η_c are explained in [13]. If, for the instantaneous SIR, λ is defined as $\lambda = R^2/r^2$, while for the average SIR, S is defined as $S = \Omega_d/\Omega_c$, then the PDF of instantaneous SIR can be given as

$$
f_\lambda(\lambda) = \sum_{j=0}^{\infty}\sum_{k=0}^{\infty} \frac{\begin{array}{c}\alpha\pi(\eta_d-1)^{2j}(\eta_d+1)^{2j+2\mu_d}(\eta_c-1)^{2k}(\eta_c+1)^{2k+2\mu_c}\\ \times\mu_c^{2k+2\mu_c}\mu_d^{2j+2\mu_d}\Gamma(2\mu_d+2\mu_c+2k+2j)\end{array}}{2^{2j+2k-1}\Gamma(\mu_d)\Gamma(\mu_c)\Gamma(\mu_d+j+1/2)\Gamma(\mu_c+k+1/2)j!k!}
$$
$$
\times\frac{\lambda^{(\alpha(2j+2\mu_d)/2)-1}S^{2k+2\mu_c}\eta_d^{2k+2\mu_c+\mu_d}\eta_c^{2j+2\mu_d+\mu_c}}{\left(\mu_d(1+\eta_d)^2\eta_c\lambda^{\alpha/2}+\mu_c(1+\eta_c)^2\eta_d S\right)^{2\mu_d+2\mu_c+2k+2j}}. \qquad (5.29)
$$

The double infinity sum in (5.29) converges rapidly, since only about 20–30 terms should be summed to achieve the accuracy at the 5th significant digit for various values of corresponding system parameters. The PDF of instantaneous SIR for various values of system parameters is presented in Figure 5.6.

Capitalizing on (5.29), the closed-form expression for the CDF of the instantaneous SIR can be presented as

$$
F_\lambda(\lambda) = \int_0^\lambda f_\lambda(t)dt
$$
$$
= \sum_{j=0}^{\infty}\sum_{k=0}^{\infty} \frac{\begin{array}{c}\pi(\eta_d-1)^{2j}(\eta_c-1)^{2k}\eta_d^{\mu_d}\eta_c^{\mu_c}\\ \times\Gamma(2\mu_d+2\mu_c+2k+2j)\end{array}}{\begin{array}{c}2^{2j+2k-2}\Gamma(\mu_d)\Gamma(\mu_c)\Gamma(\mu_d+j+1/2)\\ \times\Gamma(\mu_c+k+1/2)j!k!(\eta_d+1)^{2j+2\mu_d}(\eta_c+1)^{2k+2\mu_c}\end{array}}
$$
$$
\times B\left(2j+2\mu_d,2k+2\mu_c,\frac{\lambda^{(\alpha(2j+2\mu_d)/2)-1}S^{2k+2\mu_c}\times\eta_d^{2k+2\mu_c+\mu_d}\eta_c^{2j+2\mu_d+\mu_c}}{\left(\mu_d(1+\eta_d)^2\eta_c\lambda^{\alpha/2}+\mu_c(1+\eta_c)^2\eta_d S\right)^{2\mu_d+2\mu_c+2k+2j}}\right).
$$
$$
\qquad (5.30)
$$

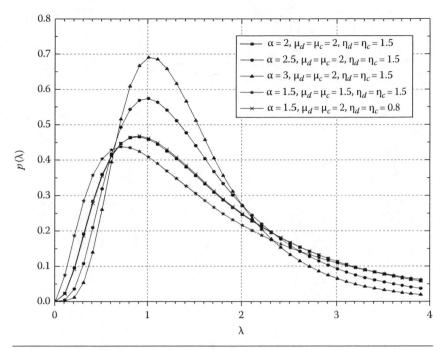

Figure 5.6 PDF of instantaneous SIR for various values of system parameters.

OP has been efficiently evaluated for various values of transmission parameters and graphically presented in Figure 5.7.

The general conclusion deduced from Figure 5.7 is that higher OP values are achieved in the areas where parameters η, μ, and α obtain higher values. Performance improvement obtained by the usage of dual-branch SC diversity is visible, since significantly lower OP values are achieved for the same system parameter values.

5.6 Performance Analysis of the Reception over Generalized K Fading Channels in the Presence of CCI

Composite propagation environments of multipath fading, super-imposed by lognormal or gamma shadowing, result in lognormal or gamma-based fading models, such as Rayleigh-, Rician-, or Nakagami-lognormal and Nakagami-gamma fading channels. However, as based fading models are analytically very difficult to handle and, therefore, rather complicated, mathematical expressions have been derived for the performance of digital communication

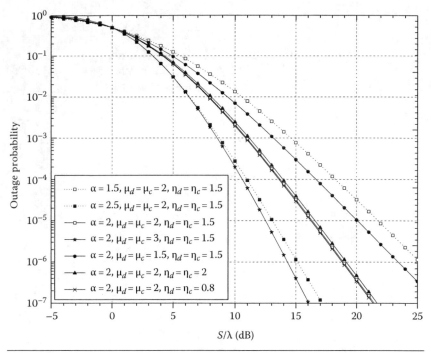

Figure 5.7 OP for various values of communication parameters in the interference limited system.

systems operating in such environments. However, the generalized K distribution [15] is a generic and versatile distribution for the accurate modeling of a great variety of short-term fading in conjunction with long-term fading (shadowing) channel conditions. This composite distribution is convenient for modeling multipath/shadowing correlated fading environments when the correlations between the instantaneous signal powers and the ones between their average powers are different. The case when transmission over generalized K fading channels has been exposed to the influence of CCI will be observed here. The desired signal envelope and interfering one are both generalized K random variables with the PDFs, respectively [16, Eq. (1)]:

$$f_R(R) = \frac{4}{\Gamma(m_d)\Gamma(k_d)}\left(\frac{m_d}{\Omega_d}\right)^{(m_d+k_d)/2} R^{m_d+k_d-1} K_{k_d-m_d}\left(2R\sqrt{\frac{m_d}{\Omega_d}}\right);$$

$$(5.31)$$

and

$$f_r(r) = \frac{4}{\Gamma(m_c)\Gamma(k_c)} \left(\frac{m_c}{\Omega_c}\right)^{(m_c+k_c)/2} r^{m_c+k_c-1} K_{k_c-m_c}\left(2r\sqrt{\frac{m_c}{\Omega_c}}\right), \qquad (5.32)$$

where $K_{k-m}(.)$ is the $(k-m)$-th-order-modified Bessel function of the second kind [9, Eq. (8.407)] and $\Omega = E(x^2)/k$. The parameter m is a fading severity parameter and k is a shadowing severity parameter. The two shaping parameters, m and k, can take different values ($m \geq 1/2$ and $k \in (0,\infty)$), where $m_d \geq 1/2$ and $m_c \geq 1/2$ represent the Nakagami-m shaping parameters for the desired and interference signal, respectively; $k_d > 0$ and $k_c > 0$ denote the shadowing shaping parameters of desired and interference signal, respectively, which approximate several shadowing conditions, from severe shadowing ($k_d, k_c \to 0$) to no shadowing ($k_d, k_c \to \infty$); Therefore, a great variety of short-term and long-term fading (shadowing) conditions can be described.

If, for the instantaneous SIR, z, is defined as $z = R^2/r^2$, while for the average SIR, S, is defined as $S = \Omega_d/\Omega_c$, then the PDF of instantaneous SIR can be given, according to [17, Eq. (07.34.21.0011.01)] as

$$f_z(z) = \left(\frac{m_d}{m_c S}\right)^{(m_d+k_d)/2} \frac{z^{(m_d+k_d)/2-1}}{\Gamma(m_d)\Gamma(k_d)\Gamma(m_c)\Gamma(k_c)} \int_0^\infty t^{(m_d+k_d+m_c+k_c)/2-1}$$

$$\times G_{0,2}^{2,0}\left[t \left|\begin{matrix} - \\ (k_c-m_c)/2, -(k_c-m_c)/2 \end{matrix}\right.\right]$$

$$\times G_{0,2}^{2,0}\left[m_d t/m_c S \left|\begin{matrix} - \\ (k_d-m_d)/2, -(k_d-m_d)/2 \end{matrix}\right.\right], \qquad (5.33)$$

where $G_{p,q}^{m,n}\left[.\left|.\right.\right]$ is the Meijer's G-function [9, Eq. (9.301[7])].

Now, starting from (5.33), the closed-form expression for the CDF of the instantaneous SIR can be obtained by using [18, Eq. (26)] in the form of

$$F_z(z) = \int_0^z f_z(u)\,du =$$

$$= \frac{z^{(m_d + k_d)/2}}{\Gamma(m_d)\Gamma(k_d)\Gamma(m_c)\Gamma(k_c)}(m_d/m_c S)^{(m_d + k_d)/2}$$

$$\times G_{3,3}^{2,3}\left[m_d z/m_c S \left|\begin{array}{l} 1-(m_d + k_d + 2k_c)/2, \\ 1-(m_d + k_d + 2m_c)/2, \\ 1-(m_d + k_d)/2(k_d - m_d)/2, \\ -(k_d - m_d)/2, -(m_d + k_d)/2 \end{array}\right.\right].$$

$$(5.34)$$

Capitalizing on expressions (5.33) and (5.34), the standard performance criteria for wireless transmission could be efficiently determined. In addition, capitalizing on these expressions, more complex cases of transmission and multichannel reception will be considered in the next chapter, where useful system analysis in function of various transmission and reception parameters will also be provided.

References

1. Panic, S. R. (2010). Mitigating the influence of α–μ multipath fading on wireless telecommunication system performances. PhD thesis, University of Nis, Nis, Serbia.
2. Stefanovic, M. et al. (2012). The CCI effect on system performance in κ-μ fading channels. *Technics Technologies Education Management*, 7(1), 88–92.
3. Matovic, M. et al. (2012). η-μ Modeled multipath propagation of electromagnetic waves. *Technics Technologies Education Management*, 7(2), 456–461.
4. Spalevic, P. et al. (2011). The co-channel interference effect on average error rates in Hoyt fading channels. *Revue Roumaine des Sciences Techniques. Series Electrotechnique et Energetique*, 56(3), 305–313.
5. Stamenkovic, G. et al. Performance analysis of wireless communication system in general fading environment subjected to shadowing and interference. *International Journal of Electronics and Communications (AEUE)*. Submitted.

6. Abu-Dayya, A. A. and Beaulieu, N. C. (1999). Diversity MPSK receivers in co-channel interference. *IEEE Transactions on Vehicular Technology*, 48(6), 1959–1965.

7. Karagiannidis, G. K. (2003). Performance analysis of SIR-based dual selection diversity over correlated Nakagami-*m* fading channels. *IEEE Transactions on Vehicular Technology*, 52(5), 1207–1216.

8. Yacoub, M. D. (2002). The α-μ distribution: A general fading distribution. *Proceedings of the 13th International Symposium on Personal, Indoor, and Mobile Radio Communications*, Vol. 2, pp. 629–633.

9. Gradshteyn, I. and Ryzhik, I. (1980). *Tables of Integrals, Series, and Products*. Academic Press, New York.

10. Milisic, M., Hamza, M., and Hadzialic, M. (2009). BEP/SEP and outage performance analysis of L-branch maximal-ratio combiner for κ-μ fading. *International Journal of Digital Multimedia Broadcasting*, 2009, Article ID 573404, 1–8.

11. Cotton, S., Scanlon, W., and Guy, J. (2008). The distribution applied to the analysis of fading in body to body communication channels for fire and rescue personnel. *IEEE Antennas and Wireless Propagation Letters*, 7, 66–69.

12. Simon, M. K. and Alouini, M. S. (2005). *Digital Communications over Fading Channels*, 2nd edn. Wiley, New York.

13. Yacoub, M. D. (2005). The symmetrical η-μ distribution: A general fading distribution. *IEEE Transactions on Broadcasting*, 51(4), 504–511.

14. Fraidenraich, G. and Yacoub, M. D. (2006). The α-η-μ and α-κ-μ fading distributions. *Proceedings of IEEE International Symposium on Spread Spectrum Techniques and Applications*, pp. 16–20.

15. Shankar, P. M. (2004). Error rates in generalized shadowed fading channels. *Wireless Personal Communications*, 28(4), 233–238.

16. Bithas, P. S., Mathiopoulos, P. T., and Kotsopoulos, S. A. (2007). Diversity reception over Generalized-*K* (*K_G*) fading channels. *IEEE Transactions on Communications*, 6(12), 4238–4243.

17. Wolfram Research, Inc. http:/functions.wolfram.com (accessed June 2012).

18. Adamchik, V. S. and Marichev, O. I. (1990). The algorithm for calculating integrals of hypergeometric type functions and its realization in reduce system. *Proceedings of International Conference on Symbolic and Algebraic Computation*, Tokyo, Japan, pp. 212–224.

6

MULTICHANNEL RECEIVER OVER FADING CHANNELS IN THE PRESENCE OF CCI

Signal performance improvement at the receiver achieved by using diverse reception techniques will be considered through the standard performance criteria, defined in Chapter 4, such as average symbol error rate (ASEP) and outage probability (OP). The usage of space diversity techniques, described in Chapter 4, will be discussed. For determining the previously mentioned criteria, it is necessary to determine joint statistics at the diversity structure inputs. Since both multipath fading and cochannel interference (CCI) are considered to be significant drawbacks, it is necessary to observe the ratios of desired signals and corresponding cochannel interferers, at the multibranch diversity structure inputs, and their joint statistics. When a diversity system is applied on small terminals with multiple antennas, which is often a real scenario in practice, fading among the branches is correlated due to insufficient antenna spacing [1]. Therefore, it is important to understand how the correlation between received signals affects the system performances. Several correlation models have been proposed in the literature for evaluating the performances of a diversity system. In this chapter, we will discuss cases of uncorrelated multibranch reception, such as the diversity reception cases over correlated fading channels.

6.1 Diversity Reception over α-μ Fading Channels in the Presence of CCI

6.1.1 SSC Diversity Reception with Uncorrelated Branches

First, we will discuss the switch-and-stay combining (SSC) diversity technique, which has already been discussed in Chapter 4. As already mentioned, in systems where the level of the CCI is sufficiently high compared to the level of thermal noise, the SSC selects a particular branch until the signal-to-interference ratio (SIR) of that branch, λ_i, falls below a predetermined threshold (SIR-based switch-and-stay diversity) [2,3]. Then, the combiner switches to another branch and stays there regardless of the SIR value of that branch.

Let us denote the predetermined switching threshold for both the input branches with z_T. When observing uncorrelated branches at the terminal, the probability density function (PDF) of the SIR, λ, at the combiner's output can be presented in the form of Equation 4.33:

$$
f_{SSC}(\lambda) = \begin{cases}
\dfrac{F_{\lambda_1}(z_T)F_{\lambda_2}(z_T)}{F_{\lambda_1}(z_T)+F_{\lambda_2}(z_T)}\left(f_{\lambda_1}(\lambda)+f_{\lambda_2}(\lambda)\right) & \lambda \le z_T; \\[4mm]
\dfrac{F_{\lambda_1}(z_T)F_{\lambda_2}(z_T)}{F_{\lambda_1}(z_T)+F_{\lambda_2}(z_T)}\left(f_{\lambda_1}(\lambda)+f_{R_2}(\lambda)\right) & \\[4mm]
\quad + \dfrac{f_{\lambda_1}(\lambda)F_{\lambda_2}(z_T)+F_{\lambda_1}(z_T)f_{\lambda_2}(\lambda)}{F_{\lambda_1}(z_T)+F_{\lambda_2}(z_T)} & \lambda > z_T,
\end{cases}
$$

$$(6.1)$$

with $F_\lambda(z_T)$ defined with

$$
F_{\lambda_i}(z_T) = \int_0^{z_T} f_{\lambda_i}(\lambda)\,d\lambda. \tag{6.2}
$$

The PDF and cumulative distribution function (CDF) of the output SIR can now be presented based on (5.6) in the following forms:

$$
f_{\lambda_1}(\lambda) = \frac{\alpha_1\mu_{d1}{}^{\mu_{d1}}\mu_{c1}{}^{\mu_{c1}}\lambda^{(\alpha_1\mu_{d1}/2)-1}}{2\left(\mu_{c1}S_1{}^{\alpha_1/2}+\mu_{d1}\lambda^{\alpha_1/2}\right)^{\mu_{d1}+\mu_{c1}}}\frac{\Gamma\left(\mu_{d1}+\mu_{c1}\right)}{\Gamma\left(\mu_{d1}\right)\Gamma\left(\mu_{c1}\right)};
$$

$$(6.3)$$

$$
f_{\lambda_2}(\lambda) = \frac{\alpha_2\mu_{d2}{}^{\mu_{d2}}\mu_{c2}{}^{\mu_{c2}}\lambda^{(\alpha_2\mu_{d2}/2)-1}}{2\left(\mu_{c2}S_1{}^{\alpha_2/2}+\mu_{d2}\lambda^{\alpha_2/2}\right)^{\mu_{d2}+\mu_{c2}}}\frac{\Gamma\left(\mu_{d2}+\mu_{c2}\right)}{\Gamma\left(\mu_{d2}\right)\Gamma\left(\mu_{c2}\right)}
$$

and

$$F_{\lambda_1}(z_T) = \int_0^{z_T} f_{\lambda_1}(\lambda)d\lambda$$

$$= \frac{\Gamma(\mu_{d1}+\mu_{c1})}{\Gamma(\mu_{d1})\Gamma(\mu_{c1})} \frac{\left(\mu_{d1}z_T{}^{\alpha_1/2}/(\mu_{c1}S_1{}^{\alpha_1/2}+\mu_{d1}z_T{}^{\alpha_1/2})\right)^{\mu_{d1}}}{\mu_{d1}}$$

$$\times {}_2F_1\left(\mu_{d1},1-\mu_{c1},1+\mu_{d1},\mu_{d1}z_T{}^{\alpha_1/2}/(\mu_{c1}S_2{}^{\alpha_1/2}+\mu_{d1}z_T{}^{\alpha_1/2})\right);$$

$$F_{\lambda_2}(z_T) = \int_0^{z_T} f_{\lambda_2}(\lambda)d\lambda$$

$$= \frac{\Gamma(\mu_{d2}+\mu_{c2})}{\Gamma(\mu_{d2})\Gamma(\mu_{c2})} \frac{\left(\mu_{d2}z_T{}^{(\alpha_2/2)}/(\mu_{c2}S_2{}^{(\alpha_2/2)}+\mu_{d2}z_T{}^{(\alpha_2/2)})\right)^{\mu_{d2}}}{\mu_{d2}}$$

$$\times {}_2F_1\left(\mu_{d2},1-\mu_{c2},1+\mu_{d2},\mu_{d2}z_T{}^{\alpha_2/2}/(\mu_{c2}S_2{}^{\alpha_2/2}+\mu_{d2}z_T{}^{\alpha_2/2})\right),$$

$$(6.4)$$

respectively. Here, $\Gamma(x)$ stands for the gamma function [4, Eq. (8.310.1)], and ${}_2F_1(u_1,u_2;u_3;x)$, the Gaussian hypergeometric function [4, Eq. (9.14)], while $S_k = \hat{R}_k^2/\hat{r}_k^2$ is the average SIRs at the kth input branch of the dual-branch SSC. Similarly, the CDF at the output of the SSC combiner can be determined by using Equation 4.33:

$$F_{SSC}(\lambda) = \begin{cases} \dfrac{F_{\lambda_1}(z_T)F_{\lambda_2}(z_T)}{F_{\lambda_1}(z_T)+F_{\lambda_2}(z_T)}(F_{\lambda_1}(\lambda)+F_{\lambda_2}(\lambda)) & \lambda \leq z_T \\[4mm] \dfrac{F_{\lambda_1}(z_T)F_{\lambda_2}(z_T)}{F_{\lambda_1}(z_T)+F_{\lambda_2}(z_T)}(F_{\lambda_1}(\lambda)+F_{\lambda_2}(\lambda)-2) \\[4mm] \quad + \dfrac{F_{\lambda_1}(\lambda)F_{\lambda_2}(z_T)+F_{\lambda_1}(z_T)F_{\lambda_2}(\lambda)}{F_{\lambda_1}(z_T)+F_{\lambda_2}(z_T)}. & \lambda > z_T \end{cases}$$

$$(6.5)$$

Now, by using the definition of OP from Section 4.1, the OP for the observed case is presented in Figure 6.1. It has been assumed that due to propagation conditions $\mu_{d1}=\mu_{d2}=\mu_d$ and $\mu_{c1}=\mu_{c2}=\mu_c$.

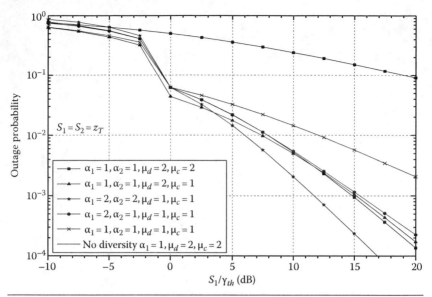

Figure 6.1 Outage probability improvement by using SSC diversity reception with uncorrelated branches for various values of system parameters.

It can be seen from Figure 6.1 that by applying the SSC diversity reception with uncorrelated branches, a significant improvement of system performances is achieved over reception with no diversity in an interference-limited environment. The improvement is visible in the wide range of received normalized SIR and for various values of system parameters.

In a similar manner, capitalizing on previous relations, another performance measure, ASEP, can be efficiently evaluated. The ASEP for the considered scenario, over BDPSK and NCFSK modulation schemes, is presented in Figures 6.2 and 6.3. It can be seen that a higher level of performance improvement is achieved in the areas where parameters α_1, α_2, and μ_d reach higher values, and when parameter μ_c decreases. Also, it is shown that using the BDPSK modulation technique provides better results than using the NCFSK.

6.1.2 SSC Diversity Reception with Correlated Branches

Now, we will consider a case when, due to insufficient antennae spacing in both desired and interfering signal, envelopes experience correlative α-μ fading with joint distributions, respectively [5]:

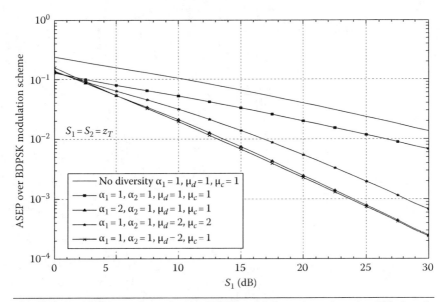

Figure 6.2 ASEP improvement for the BDPSK modulation scheme by using SSC diversity reception with uncorrelated branches for various values of system parameters.

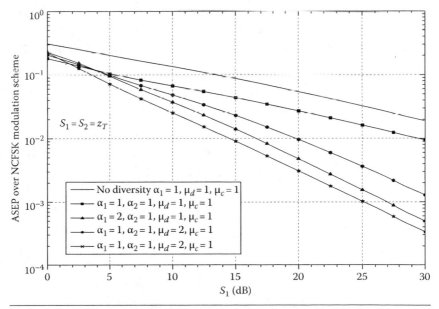

Figure 6.3 ASEP improvement for the NCFSK modulation scheme by using SSC diversity reception with uncorrelated branches for various values of system parameters.

$$f_{R_1,R_2}\left(R_1,R_2\right)=\frac{\alpha_1\mu_d^{\mu_d}R_1^{\alpha_1\mu_d-1}}{\hat{R}_1^{\alpha_1\mu_d}\Gamma\left(\mu_d\right)}$$

$$\times\exp\left(-\frac{\mu_dR_1^{\alpha_1}}{\hat{R}_1^{\alpha_1}}\right)\frac{\alpha_2\mu_d^{\mu_d}R_2^{\alpha_2\mu_d-1}}{\hat{R}_2^{\alpha_2\mu_d}\Gamma\left(\mu_d\right)}\exp\left(-\frac{\mu_dR_2^{\alpha_2}}{\hat{R}_2^{\alpha_2}}\right)$$

$$\times\sum_{l=0}^{\infty}\frac{l!\Gamma\left(\mu_d\right)}{\Gamma\left(\mu_d+l\right)}\rho_{12d}^{l}L_l^{\mu_d-1}\left(\frac{\mu_dR_1^{\alpha_1}}{\hat{R}_1^{\alpha_1}}\right)L_l^{\mu_d-1}\left(\frac{\mu_dR_2^{\alpha_2}}{\hat{R}_2^{\alpha_2}}\right); \tag{6.6}$$

$$f_{r_1,r_2}\left(r_1,r_2\right)=\frac{\alpha_1\mu_c^{\mu_c}r_1^{\alpha_1\mu_c-1}}{\hat{r}_1^{\alpha_1\mu_c}\Gamma\left(\mu_c\right)}\exp\left(-\frac{\mu_cr_1^{\alpha_1}}{\hat{r}_1^{\alpha_1}}\right)\frac{\alpha_2\mu_c^{\mu_c}r_2^{\alpha_2\mu_c-1}}{\hat{r}_2^{\alpha_2\mu_c}\Gamma\left(\mu_c\right)}\exp\left(-\frac{\mu_cr_2^{\alpha_2}}{\hat{r}_2^{\alpha_2}}\right)$$

$$\times\sum_{k=0}^{\infty}\frac{k!\Gamma\left(\mu_c\right)}{\Gamma\left(\mu_c+k\right)}\rho_{12c}^{k}L_k^{\mu_c-1}\left(\frac{\mu_cr_1^{\alpha_1}}{\hat{r}_1^{\alpha_1}}\right)L_k^{\mu_c-1}\left(\frac{\mu_cr_2^{\alpha_2}}{\hat{r}_2^{\alpha_2}}\right). \tag{6.7}$$

Here, it is important to quote that ρ_{12d} and ρ_{12c} denote the power and interfering signal correlation coefficients defined as $cov(R_i^2,R_j^2)/(var(R_i^2)var(R_j^2))^{1/2}$, and $cov(r_i^2,r_j^2)/(var(r_i^2)var(r_j^2))^{1/2}$, respectively; $\hat{R}_i=\sqrt[\alpha_i]{E\left(R_i^{\alpha_i}\right)}$ and $\hat{r}_i=\sqrt[\alpha_i]{E\left(r_i^{\alpha_i}\right)}$ denote the α-root mean values of process, while $L_n^k(x)$ is a generalized Laguerre polynomial given by [6]. $E(.)$ stands for operator of mathematical expectation.

Let $z_1=R_1^2/r_1^2$ and $z_2=R_2^2/r_2^2$ represent the instantaneous SIR on the diversity branches, respectively. The joint PDF of z_1 and z_2 can be expressed by [7]

$$f_{z_1,z_2}\left(z_1,z_2\right)=\frac{1}{4\sqrt{z_1z_2}}\int_0^\infty\int_0^\infty f_{R_1,R_2}\left(r_1\sqrt{z_1},r_2\sqrt{z_2}\right)f_{r_1,r_2}\left(r_1,r_2\right)r_1r_2dr_1dr_2.$$

$$\tag{6.8}$$

By substituting Equations 6.6 and 6.7 in (6.8), we obtain

$$
f_{z_1 z_2}(z_1, z_2) = \sum_{l=0}^{\infty} \sum_{k=0}^{\infty} \sum_{j=0}^{l} \sum_{m=0}^{l} \sum_{n=0}^{k}
$$

$$
\times \sum_{p=0}^{k} \frac{G_1 z_1^{(\alpha_1(\mu_d + j)/2)-1} z_2^{(\alpha_2(\mu_d + m)/2)-1}}{\left(\mu_d z_1^{\alpha_1/2} + S_1^{\alpha_1/2}\right)^{\mu_d + \mu_c + j + n} \left(\mu_c z_2^{\alpha_2/2} + S_2^{\alpha_2/2}\right)^{\mu_d + \mu_c + m + p}}, \quad (6.9)
$$

with $S_k = \hat{R}_k^2 / \hat{r}_k^2$ being the average SIRs at the kth input branch of the dual-branch selection combining (SC) and

$$
G_1 = \frac{4(-1)^{j+m+n+p} \alpha_1 \alpha_2 \mu_d^{2\mu_d + j + m} \mu_c^{2\mu_c + n + p} l!\, k!\, \rho_{12d}^k \rho_{12c}^l}{j!\, m!\, n!\, p!\, (l-j)!\, (l-m)!\, (k-n)!\, (k-p)!}
$$

$$
\times \frac{\Gamma(\mu_d + \mu_c + n + j)\Gamma(\mu_d + \mu_c + p + m)}{\Gamma(\mu_d)\Gamma(\mu_c)\Gamma(\mu_d + l)\Gamma(\mu_c + k)}
$$

$$
\times \frac{\left((\mu_d - 1 + l)!\right)^2 \left((\mu_c - 1 + k)!\right)^2 S_1^{(\alpha_1(n+\mu_d))/2} S_2^{(\alpha_2(p+\mu_c))/2}}{(\mu_d - 1 + j)!\,(\mu_d - 1 + m)!\,(\mu_c - 1 + n)!\,(\mu_c - 1 + p)!}.
$$

$$(6.10)$$

Let z represent the instantaneous SIR at the SSC output, and z_T the predetermined switching threshold for both the input branches. Following [8], the PDF of z is given by

$$
f_{SSC}(z) = \begin{cases} v_{SSC}(z) & z \leq z_T; \\ v_{SSC}(z) + f_{z_1}(z) & z > z_T, \end{cases} \quad (6.11)
$$

where $v_{SSC}(z)$, according to [8], can be expressed as

$$
v_{SSC}(z) = \int_0^{z_T} f_{z_1, z_2}(z, z_2)\, dz_2. \quad (6.12)
$$

Moreover, $v_{SSC}(z)$ can be expressed as an infinite series:

$$v_{SSC}(z) = \sum_{l=0}^{\infty} \sum_{k=0}^{\infty} \sum_{j=0}^{l} \sum_{m=0}^{l} \sum_{n=0}^{k}$$

$$\times \sum_{p=0}^{k} G_2 \frac{z^{(\alpha_1(\mu_d+j)/2)-1} {}_2F_1\left(\mu_c+m, 1-\mu_d+p, \mu_c+m+1, G_3\right)}{\left(\mu_c S_1^{\alpha_1/2} + \mu_d z^{\alpha_1/2}\right)^{\mu_d+\mu_c+j+n}} G_3^{\mu_c+m},$$

$$(6.13)$$

with

$$G_2 = \frac{(-1)^{j+m+n+p} \alpha_1 \mu_d^{\mu_d+j} \mu_c^{\mu_c+n} l! k! \rho_{12d}^k \rho_{12c}^l}{2 j! m! n! p! (l-j)! (l-m)! (k-n)! (k-p)!}$$

$$\times \frac{\Gamma(\mu_d+\mu_c+n+j) \Gamma(\mu_d+\mu_c+p+m)}{\Gamma(\mu_d) \Gamma(\mu_c) \Gamma(\mu_d+l) \Gamma(\mu_c+k)(\mu_c+m)}$$

$$\times \frac{\left((\mu_d-1+l)!\right)^2 \left((\mu_c-1+k)!\right)^2 S_1^{(\alpha_1(n+\mu_d))/2}}{(\mu_d-1+j)! (\mu_d-1+m)! (\mu_c-1+n)! (\mu_c-1+p)!};$$

$$G_3 = \frac{\mu_d z_T^{\alpha_2/2}}{\mu_c S_2^{\alpha_2/2} + \mu_d z_T^{\alpha_2/2}}. \qquad (6.14)$$

Assuming that due to propagation conditions, $\mu_{d1}=\mu_{d2}=\mu_d$ and $\mu_{c1}=\mu_{c2}=\mu_c$, in the same manner, the $f_{z_1}(z)$ can be expressed as

$$f_{z_1}(z) = \frac{\alpha_1 \mu_d^{\mu_d} \mu_c^{\mu_c} z^{(\alpha_1\mu_d/2)-1}}{2\left(\mu_c S_1^{\alpha_1/2} + \mu_d z^{\alpha_1/2}\right)^{\mu_d+\mu_c}} \frac{\Gamma(\mu_d+\mu_c)}{\Gamma(\mu_d) \Gamma(\mu_c)}. \qquad (6.15)$$

Similar to (6.11), the CDF of the SSC output SIR, that is, the $F_{z_{SSC}}(z)$, is given by [9]

$$F_{z_{SSC}}(z) = \begin{cases} F_{z_1,z_2}(z,z_\tau), & z \leq z_\tau, \\ F_z(z) - F_z(z_\tau) + F_{z_1,z_2}(z,z_\tau), & z > z_\tau. \end{cases} \qquad (6.16)$$

By substituting (6.9) in $F_{z_1,z_2}(z,z_\tau) = \int_0^z \int_0^{z_\tau} f_{z_1,z_2}(z_1,z_2) dz_1 dz_2$ and

(6.15) in $F_z(z) = \int_0^z f_z(z) dz$, $F_{z_1,z_2}(z,z_\tau)$ and $F_z(z)$ can be expressed as the following infinite series, respectively:

$$F_{z_1,z_2}(z,z_\tau) = \sum_{l=0}^{\infty}\sum_{k=0}^{\infty}\sum_{j=0}^{l}\sum_{m=0}^{l}\sum_{n=0}^{k}\sum_{p=0}^{k} G_4 G_5^{\mu_1+l} G_6^{\mu_1+m}$$

$$\times\ _2F_1\left(\mu_d+j,1-\mu_c-n;1+\mu_d+j;G_5\right)$$

$$\times\ _2F_1\left(\mu_d+m,1-\mu_c-p;1+\mu_d+m;G_3\right);$$

$$F_z(z) = \frac{\Gamma(\mu_d+\mu_c)}{\Gamma(\mu_d)\Gamma(\mu_c)}\frac{(G_5)^{\mu_d}}{\mu_d}\ _2F_1\left(\mu_d,1-\mu_c,1+\mu_d,G_5\right); \qquad (6.17)$$

$$F_z(z_T) = \frac{\Gamma(\mu_d+\mu_c)}{\Gamma(\mu_d)\Gamma(\mu_c)}\frac{(G_6)^{\mu_d}}{\mu_d}\ _2F_1\left(\mu_d,1-\mu_c,1+\mu_d,G_6\right),$$

with

$$G_4 = \frac{(-1)^{j+m+n+p}\ l!k!\rho_{12d}^k\rho_{12c}^l}{j!n!m!\ p!(l-j)!(l-m)!(\mu_d+l)(\mu_c+m)}$$

$$\times\frac{\Gamma(\mu_d+\mu_c+n+j)\Gamma(\mu_d+\mu_c+p+m)}{\Gamma(\mu_d)\Gamma(\mu_c)\Gamma(\mu_d+l)\Gamma(\mu_c+k)}$$

$$\times\frac{\mu_d^{2\mu_d+j+p}\mu_c^{2\mu_c+n+m}\left((\mu_d-1+l)!\right)^2\left((\mu_c-1+k)!\right)^2}{(k-n)!(k-p)!(\mu_d-1+j)!(\mu_d-1+m)!};$$
$$(\mu_c-1+n)!(\mu_c-1+p)!$$

$$G_5 = \frac{\mu_d z^{\alpha_1/2}}{\mu_c S_1^{\alpha_1/2}+\mu_d z^{\alpha_1/2}}; \quad G_6 = \frac{\mu_d z_T^{\alpha_2/2}}{\mu_c S_2^{\alpha_2/2}+\mu_d z_T^{\alpha_2/2}}. \qquad (6.18)$$

The number of terms needed to be summed in (6.17) to achieve accuracy at the desired significant digit is as depicted in Table 6.1. The terms need to be summed for achieving a desired accuracy and this depends strongly on the correlation coefficients, ρ_{d12} and ρ_{c12}. It is obvious that the number of the terms increases as correlation coefficients increase. Also, when $\rho_{c12} > \rho_{d12}$, we need more terms for correct computation.

OP performance results have been obtained using (6.17). These results are presented in Figure 6.4, as the function of the normalized outage threshold for several values of ρ_{d12}, ρ_{c12}, μ_d, μ_c, α_1, and α_2.

Table 6.1 Terms That Need to Be Summed in (6.17) to Achieve Accuracy at the 6th Significant Digit

$S_1/Z = 10$ dB, $\alpha_1 = \alpha_2 = 2$ $S_1 = S_2 = Z_T$		$\mu_d = 1$ $\mu_c = 1$	$\mu_d = 1.2$ $\mu_c = 1.5$
$\rho_{d12} = 0.3$	$\rho_{c12} = 0.2$	24	21
$\rho_{d12} = 0.3$	$\rho_{c12} = 0.3$	28	25
$\rho_{d12} = 0.3$	$\rho_{c12} = 0.4$	37	35
$\rho_{d12} = 0.4$	$\rho_{c12} = 0.3$	31	27
$\rho_{d12} = 0.5$	$\rho_{c12} = 0.5$	51	47

Note: We consider a dual- and triple-branch selection combining diversity system.

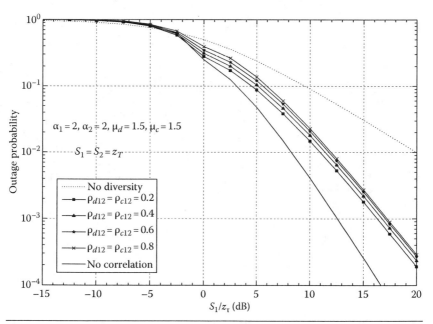

Figure 6.4 OP improvement by using SSC diversity reception with correlated branches for various values of correlation level.

Normalized outage threshold is defined as average SIR at the input branch S_1 of the balanced $(S_1 = S_2)$ dual-branch switch-and-stay combiner divided by the specified threshold value z_τ. Results show that as the signal and interference correlation coefficients ρ_{d12} and ρ_{c12} increase, and as normalized outage threshold decreases, the OP increases.

The ASEP, $\bar{P}_e$, can be evaluated by averaging the conditional symbol error probability (SEP) for a given SIR, that is, $P_e(z)$, over the PDF of z_{SSC}, that is, $f_{zSSC}(z)$ [3]:

$$\bar{P}_e = \int_0^\infty P_e(z) f_{zSSC}(z) dz, \qquad (6.19)$$

where $P_e(z)$ depends on the applied modulation scheme. Conditional SEP for a given SIR threshold can be expressed as $P_e(z) = 1/2 \exp[-\lambda z]$, where $\lambda = 1$ for binary DPSK and $\lambda = 1/2$ for NCFSK. Hence, substituting (6.13) and (6.15) into (6.19) gives the following ASEP expression for the considered dual-branch SSC receiver:

$$\bar{P}_e = \int_0^\infty P_e(z) v_{SSC}(z) dz + \int_{z_\tau}^\infty P_e(z) f_z(z) dz. \qquad (6.20)$$

Using the previously derived infinite series expressions, we present representative numerical performance evaluation results of the studied dual-branch SSC diversity receiver, such as OP and ASEP, in the case of two modulation schemes: BDPSK and NCFSK. Applying (6.20) on BDPSK and NCFSK modulation schemes, the ASEP performance results have been obtained as a function of the average SIRs at the input branches of the balanced dual-branch SSC, that is, $S_1 = z_\tau$, for several values of ρ_{d12}, ρ_{c12}, μ_d, μ_c, α_1, and α_2. These results are plotted in Figures 6.5 and 6.6. It is shown that while the signal and interference correlation coefficients, ρ_{d12} and ρ_{c12}, increase and as the average SIRs at the input branches increase, the ASEP increases at the same time. It is very interesting to observe that for lower values of S_1, due to the fact that the interference is comparable to the desired signal, ASEP deteriorates more severely when the fading severity of the signal and interferers changes. So the increase in the α_1 and α_2 makes performance deterioration more visible in this area, and while the increase in α_1 and α_2 is higher, the deterioration is larger. Finally, by comparison of Figures 6.5 and 6.6, one can notice the better performance of BDPSK modulation scheme versus NCFSK modulation scheme.

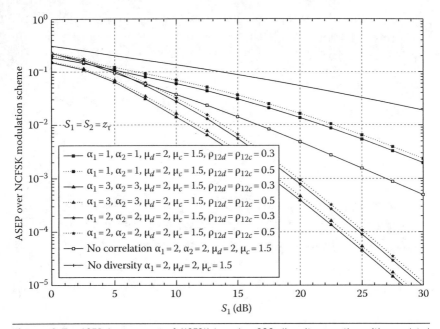

Figure 6.5 ASEP improvement of NCFSK by using SSC diversity reception with correlated branches for various values of system parameters and comparison with uncorrelated case.

6.1.3 SC Diversity Reception with Uncorrelated Branches

As explained in previous section, the SC diversity receiver selects and outputs the branch with the highest instantaneous value of SIR:

$$\lambda = \lambda_{out} = \max (\lambda_1, \lambda_2, \ldots, \lambda_N). \tag{6.21}$$

For this case, the CDF of the SC output SIR can be determined according to [10] by equating arguments $t_1 = t_2 = \cdots = t_N$ in

$$F_\lambda(t) = F_{\lambda_1}(t) F_{\lambda_2}(t) \cdots F_{\lambda_N}(t) = \prod_{i=1}^{N} F_{\lambda_i}(t)$$

$$= \prod_{i=1}^{N} \frac{\Gamma(\mu_d + \mu_c)}{\Gamma(\mu_d)\Gamma(\mu_c)} \frac{\left(\mu_d t^{\alpha_i/2}/(\mu_c S_i^{\alpha_i/2} + \mu_d t^{\alpha_i/2})\right)^{\mu_d}}{\mu_d}$$

$$\times {}_2F_1\left(\mu_d, \mu_c, \frac{\mu_d t^{\alpha_i/2}}{\mu_c S_i^{\alpha_i/2} + \mu_d t^{\alpha_i/2}}\right). \tag{6.22}$$

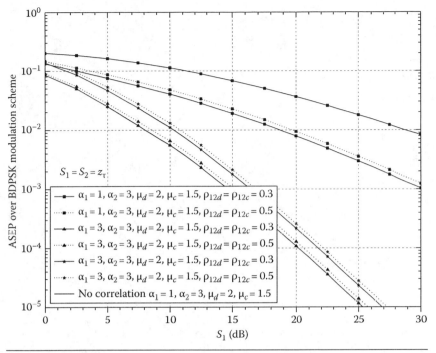

Figure 6.6 ASEP improvement of BDPSK by using SSC diversity reception with correlated branches for various values of system parameters and comparison with uncorrelated case.

Then the PDF of the SC output SIR can be presented as

$$f_\lambda(\lambda) = \frac{d}{d\lambda} F(\lambda) = \sum_{i=1}^{n} f_{\lambda_i}(\lambda) \prod_{\substack{j=1 \\ j\neq i}}^{n} F_{\lambda_i}(\lambda)$$

$$= \sum_{i=1}^{n} \frac{\alpha_i \mu_d{}^{\mu_d} \mu_c{}^{\mu_c} \lambda^{(\alpha_i \mu_d/2)-1}}{2\left(\mu_c S_i{}^{\alpha_i/2} + \mu_d \lambda^{\alpha/2}\right)^{\mu_d+\mu_c}} \frac{\Gamma(\mu_d+\mu_c)}{\Gamma(\mu_d)\Gamma(\mu_c)}$$

$$\times \prod_{\substack{j=1 \\ j\neq i}}^{n} \frac{\Gamma(\mu_d+\mu_c)}{\Gamma(\mu_d)\Gamma(\mu_c)} \frac{\left(\mu_d \lambda^{\alpha_i/2}/(\mu_c S_i{}^{\alpha_i/2} + \mu_d \lambda^{\alpha_i/2})\right)^{\mu_d}}{\mu_d}$$

$$\times {}_2F_1\left(\mu_d, \mu_c, \frac{\mu_d \lambda^{\alpha_i/2}}{\mu_c S_i{}^{\alpha_i/2} + \mu_d \lambda^{\alpha_i/2}}\right). \tag{6.23}$$

The OP for the case of SC diversity reception can be obtained using (6.22). OP values as the function of normalized threshold

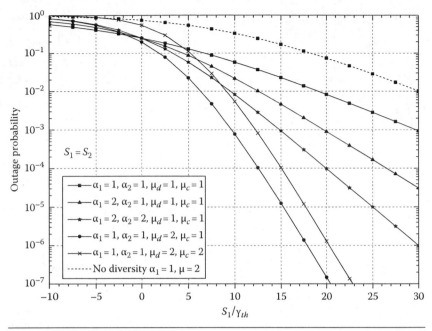

Figure 6.7   Dual-branch SC receiver OP versus normalized threshold S_1/γ for various values of system parameters.

(protection ratio) are presented in Figures 6.7 and 6.8. It is obvious from Figure 6.7 that dual-branch SC reception significantly improves OP, because lower values of this performance measure are obtained in the observed input range for all values of fading influence on deserving signal and CCI describing parameters, α_1, α_2, μ_d, and μ_c. Previous discussions about the occurrence of lower OP values remain valid for higher values of α_1, α_2, and μ_d. Since the undesirable effects of CCI are strengthening with increasing parameter μ_c, OP values are also higher in this area of higher μ_c values. In Figure 6.8, OP values are presented for the cases of SC reception with arbitrary order and for the cases that consider balanced and unbalanced reception. It is evident how significant performance improvement is obtained when a system of a higher order of diversity is applied. The highest improvement is achieved when performing dual-branch SC reception over nondiversity reception, while other improvements caused by diversity order growth are notably smaller (i.e., achieved improvement by performing triple-branch SC over SC with four branches). The effect of balanced reception (when instantaneous SIR values of input branches are equal)

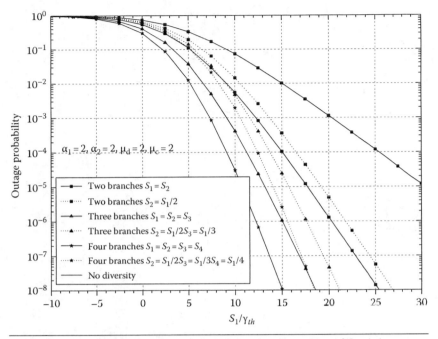

Figure 6.8 OP improvement in the function of SC diversity order and input SIR unbalance.

on OP values is also visible, that is, OP values increase when balanced reception is absent.

ASEP values of BDPSK- and NCFSK-modulated SC reception in the presence of α-μ fading and CCI are presented in Figures 6.9 and 6.10. Diversity order increase leads to performance improvement, since lower ASEP values are occurring. As expected, the highest relative improvement is achieved with dual-branch reception over nondiversity case. In addition, balanced reception improves ASEP performances. Finally, it can be noticed that the BDPSK modulation technique usage in observed conditions provides smaller ASEP values, compared to the NCFSK modulation scheme usage case.

Another valuable parameter, which serves as an excellent indicator of comprehensive performances and system reliability, is average SIR value, which can be obtained using first-order moment of SC received SIR, that is, according to relation [11]

$$\bar{\lambda} = \int_0^\infty \lambda f_\lambda(\lambda)\, d\lambda. \tag{6.24}$$

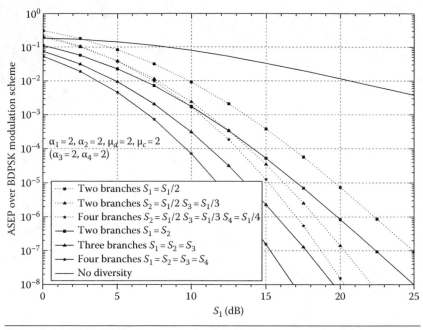

Figure 6.9   BDPSK ASEP improvement as function of SC diversity order and input SIR unbalance.

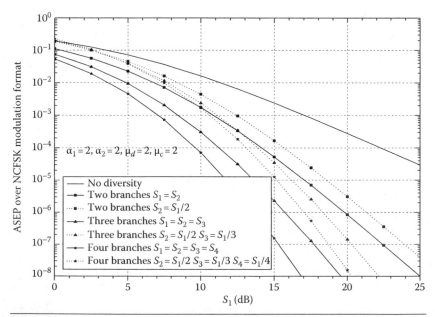

Figure 6.10   NCFSK ASEP improvement in the function of SC diversity order and input SIR unbalance.

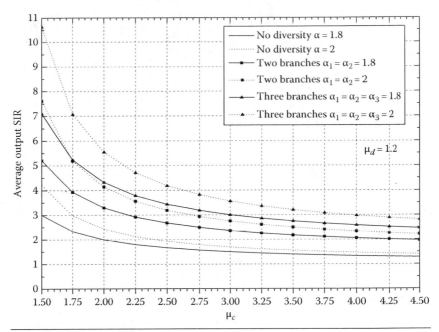

Figure 6.11   Average output SIR for arbitrary order SC reception with uncorrelated branches.

After substituting (6.23) into (6.24), average SIR value is presented in Figure 6.11, for the case of SC reception with uncorrelated branches. It is visible that the average SIR value at the reception increases with increasing diversity order. Higher μ_c values enhance undesirable CCI effect on reception characteristics, so average SIR values reach smaller values in this domain of μ_c parameter increase. The fading severity decrease (higher values of α parameter) leads to expected increase in average SIR value.

Diversity reception's ability to reduce fluctuations, caused by multipath propagation and CCI, can be determined through the parameters that consider higher-order moments of SC receiver output SIR. The amount of fading (AoF) is a parameter defined as in [12]:

$$\text{AoF} = \frac{E\left(\lambda^2\right) - \left(E\left(\lambda\right)\right)^2}{\left(E\left(\lambda\right)\right)^2} \tag{6.25}$$

and can be used for this ability evaluation.

After substituting (6.23) into (6.25), AoF values are efficiently evaluated and presented in Figure 6.12. One can observe that AoF values at reception are significantly decreasing with increasing the applied

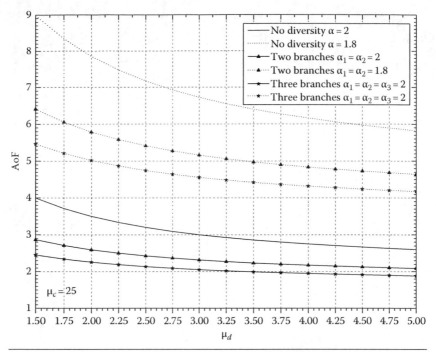

Figure 6.12   AoF improvement for arbitrary order SC reception with uncorrelated branches.

diversity order. Increase in parameter μ_d means that the number of desired signal multipath components also increases, so AoF values in this domain decrease. Decreasing fading severity, as expected, leads to a significant decrease in AoF values too.

6.1.4 SC Diversity Reception over Constantly Correlated α-μ Fading Channels

Let us now consider the constant correlation model of α-μ fading distribution. The constant correlation model can be obtained from [12] by setting $\Sigma_{i,j} \equiv 1$ for $i=j$ and $\Sigma_{i,j} \equiv \rho$ for $i \neq j$ in correlation matrix, where ρ is the correlation coefficient. The power correlation coefficient ρ_d of the desired signal is defined as $cov(R_i^2, R_j^2)/(var(R_i^2)var(R_j^2))^{1/2}$, and the power correlation coefficient ρ_c of interference is defined as $cov(r_i^2, r_j^2)/(var(r_i^2)var(r_j^2))^{1/2}$.

There is a need to derive the joint statistics for multiple α-μ variables. We are relaying on some results that are already available in the literature for the constant correlation model of Nakagami-m distribution [13].

Let

1. $R_{N_1}, \ldots, R_{N_n}$ be n Nakagami-m variates, whose marginal statistics are respectively described by the parameters $m_1 = \cdots = m_n = m$, $E(R_{N1}^2) = \Omega_1, \ldots, E(R_{Nn}^2) = \Omega_n$
2. $R_1, \ldots, R_n$ be n α-μ variates, whose marginal statistics are respectively described by the parameters $\mu_1 = \cdots = \mu_n = \mu$, α_1, $\hat{R}_1, \ldots, \alpha_n, \hat{R}_n$
3. $0 \leq \rho_{Nakagami-m} \leq 1$ be a Nakagami-m correlation parameter
4. $0 \leq \rho_{\alpha-\mu} \leq 1$ be a α-μ correlation parameter

The joint probability density function $f_{R_{N_1}, \ldots, R_{N_n}}(R_{N_1}, \ldots, R_{N_n})$ of n Nakagami-m variates $R_{N_1}, \ldots R_{N_n}$, with marginal statistics just described, is given by [13, Eq. (10)]. We use the relation between α-μ and the Nakagami-m envelopes $\left(R_{\alpha-\mu}^\alpha = R_{Nakagami-m}^2\right)$, [14, Eq. (19)], so that $R_1^{\alpha_1} = R_{N1}^2, \ldots, R_n^{\alpha_n} = R_{Nn}^2$ and we find that $\hat{R}_1^{\alpha_1} = \Omega_1, \ldots, \hat{R}_n^{\alpha_n} = \Omega_n$ and $\mu = m$. In addition, the relation between the correlation coefficient of α-μ distribution and Nakagami-m distribution have already been derived [14, Eq. (31)]. For two Nakagami-m and α-μ variates, it can be seen that $\rho_{\alpha-\mu} = \rho_{Nakagami-m}\sqrt{\mu_1/\mu_2}$. Now, because in this case we have $\mu_1 = \mu_2 = \mu_d$ for desired signal and $\mu_1 = \mu_2 = \mu_c$ for interference signal, then correlation coefficients of α-μ distribution are equal to the correlation coefficients of Nakagami-m distribution for both desired and interference signal. Finally, considering all proposed relations and [13, Eq. (10)], the joint PDF $f_{R_1, \ldots, R_n}(R_1, \ldots, R_n)$ of n α-μ variates $R_1, \ldots, R_n$ is found as [15]

$$f_{R_1,\ldots,R_n}(R_1,\ldots,R_n) = |J| f_{R_{N_1},\ldots,R_{N_n}}(R_{N_1},\ldots,R_{N_n}), \qquad (6.26)$$

with J as the Jacobian of the transformation given by

$$|J| = \begin{vmatrix} \dfrac{\partial R_{N1}}{\partial R_1} & \dfrac{\partial R_{N1}}{\partial R_2} & \cdots & \dfrac{\partial R_{N1}}{\partial R_n} \\ \dfrac{\partial R_{N2}}{\partial R_1} & \dfrac{\partial R_{N2}}{\partial R_2} & \cdots & \dfrac{\partial R_{N2}}{\partial R_n} \\ \vdots & \vdots & \vdots & \vdots \\ \dfrac{\partial R_{Nn}}{\partial R_1} & \dfrac{\partial R_{Nn}}{\partial R_2} & \cdots & \dfrac{\partial R_{Nn}}{\partial R_n} \end{vmatrix}$$

$$= \frac{\alpha_1 \alpha_2 \cdots \alpha_n}{2^n} R_1^{(\alpha_1/2)-1} R_2^{(\alpha_2/2)-1} \cdots R_n^{(\alpha_n/2)-1}. \qquad (6.27)$$

After the standard statistical procedure of transformation of variates and after some mathematical manipulations and simplifications, joint PDFs for both desired and interfering signal envelopes can be respectively expressed as

$$f_{R_1,\dots,R_n}(R_1,\dots,R_n)$$

$$= \frac{\left(1-\sqrt{\rho_d}\right)^{\mu_d}}{\Gamma(\mu_d)} \sum_{k_1,\dots,k_n=0}^{\infty} \frac{\Gamma(\mu_d+k_1+\cdots+k_n)\rho_d^{(k_1+\cdots+k_n)/2}}{\left(1-\sqrt{\rho_d}\right)^{n\mu_d+k_1+\cdots+k_n}}$$

$$\times \mu_d^{n\mu_d+k_1+\dots+k_n} \left(\frac{1}{1+(n-1)\sqrt{\rho_d}}\right)^{\mu_d+k_1+\cdots+k_n}$$

$$\times \prod_{i=1}^{n} \frac{\alpha_i}{\Gamma(\mu_d+k_i)k_i!\hat{R}_i^{\alpha_i(\mu_d+k_i)}} R_i^{\alpha_i(\mu_d+k_i)-1} \exp\left(-\frac{\mu_d R_i^{\alpha_i}}{\hat{R}_i^{\alpha_i}(1-\sqrt{\rho_d})}\right)$$

$$(6.28)$$

and

$$f_{r_1,\dots,r_n}(r_1,\dots,r_n)$$

$$= \frac{\left(1-\sqrt{\rho_c}\right)^{\mu_c}}{\Gamma(\mu_c)} \sum_{l_1,\dots,l_n=0}^{\infty} \frac{\Gamma(\mu_c+l_1+\dots+l_n)\rho_c^{(l_1+\cdots+l_n)/2}}{\left(1-\sqrt{\rho_c}\right)^{n\mu_c+l_1+\cdots+l_n}}$$

$$\times \mu_c^{n\mu_c+l_1+\cdots+l_n} \left(\frac{1}{1+(n-1)\sqrt{\rho_c}}\right)^{\mu_c+l_1+\cdots+l_n}$$

$$\times \prod_{i=1}^{n} \frac{\alpha_i}{\Gamma(\mu_c+l_i)l_i!\hat{r}_i^{\alpha_i(\mu_c+l_i)}} r_i^{\alpha_i(\mu_c+l_i)-1} \exp\left(-\frac{\mu_c r_i^{\alpha_i}}{\hat{r}_i^{\alpha_i}(1-\sqrt{\rho_c})}\right),$$

$$(6.29)$$

where ρ_d and ρ_c are the correlation coefficients and μ_d and μ_c are the fading severity parameters for the desired and interference signals, respectively. The average desired signal and interference powers at the k-th branch are denoted by $\hat{R}_k$ and $\hat{r}_k$, respectively. Instantaneous values of SIR at the k-th input diversity branch of SC could be defined as $\lambda_k = R_k^2/r_k^2$. Let $S_k = \hat{R}_k^2/\hat{r}_k^2$ be the average SIR at the kth input branch of the multibranch SC. The joint PDF of instantaneous values of SIRs at the n input branches of SC, λ_k, $k = 1, 2, \ldots, n$, can be given by [16]

$$f_{\lambda_1,\ldots,\lambda_n}\left(t_1,\ldots,t_n\right) = \frac{1}{2^n\sqrt{t_1\ldots t_n}} \underbrace{\int_0^\infty \int_0^\infty \cdots \int_0^\infty}_{n} f_{R_1,\ldots,R_n}\left(r_1\sqrt{t_1},\ldots,r_n\sqrt{t_n}\right)$$

$$\times f_{r_1,\ldots r_n}\left(r_1,\ldots,r_n\right) \times r_1 \ldots r_n dr_1 \ldots dr_n. \qquad (6.30)$$

Substituting (6.28) and (6.29) in (6.30), $f_{\lambda_1,\ldots,\lambda_n}\left(t_1,\ldots,t_n\right)$ can be written as

$$f_{\lambda_1,\ldots,\lambda n}\left(t_1,\ldots,t_n\right)$$

$$= \underbrace{\sum_{k_1,\ldots,k_n=0}^{\infty} \sum_{l_1,\ldots,l_n=0}^{\infty}}_{2n} \frac{\alpha_1\cdots\alpha_n}{2^n} \mu_d^{n\mu_d+k_1+\cdots+k_n}$$

$$\times \mu_c^{n\mu_c+l_1+\cdots+l_n}\left(1-\sqrt{\rho_d}\right)^{n\mu_c+l_1+\cdots+l_n}\left(1-\sqrt{\rho_c}\right)^{n\mu_d+k_1+\cdots+k_n}$$

$$\times G_7 \times \prod_{i=1}^{n} \frac{t_i^{(\alpha_i(\mu_d+k_i)/2)-1} S_i^{\alpha_i(\mu_c+l_i)/2}}{\left(\mu_d\left(1-\sqrt{\rho_c}\right)t_i^{\alpha_i/2} + \mu_c\left(1-\sqrt{\rho_d}\right)S_i^{\alpha_i/2}\right)^{\mu_d+\mu_c+k_i+l_i}},$$

$$(6.31)$$

with

$$G_7 = \frac{\left(1-\sqrt{\rho_d}\right)^{\mu_d}\left(1-\sqrt{\rho_c}\right)^{\mu_c}}{\Gamma(\mu_d)\Gamma(\mu_c)}$$

$$\times \Gamma\left(\mu_d+k_1+\cdots+k_n\right)\Gamma\left(\mu_c+l_1+\cdots+l_n\right)\rho_d^{\frac{k_1+\cdots+k_n}{2}}\rho_c^{\frac{l_1+\cdots+l_n}{2}}$$

$$\times \left(\frac{1}{1+(n-1)\sqrt{\rho_d}}\right)^{\mu_d+k_1+\cdots+k_n}\left(\frac{1}{1+(n-1)\sqrt{\rho_c}}\right)^{\mu_c+l_1+\cdots+l_n}$$

$$\times \prod_{i=1}^{n}\frac{\Gamma\left(\mu_d+\mu_c+k_i+l_i\right)}{\Gamma\left(\mu_d+k_i\right)\Gamma\left(\mu_c+l_i\right)k_i!l_i!}. \qquad (6.32)$$

For this case, the joint CDF of λ_k, $k = 1, 2, \ldots, n$, can be written as [16]

$$F_{\lambda_1,\lambda_2,\ldots,\lambda_n}(t_1,t_2,\ldots,t_n) = \underbrace{\int_0^{t_1}\int_0^{t_2}\cdots\int_0^{t_n}}_{n} f_{\lambda_1,\lambda_2,\ldots,\lambda_n}(x_1,x_2,\ldots,x_n)\,dx_1\,dx_2\cdots dx_n.$$

(6.33)

By substituting expression (6.31) in (6.33) and after n successive integrations, the joint CDF becomes

$$F_{\lambda_1,\ldots,\lambda_n}(t_1,\ldots,t_n) = \underbrace{\sum_{k_1,\ldots,k_n=0}^{\infty}\sum_{l_1,\ldots,l_n=0}^{\infty}}_{2n} G_7$$

$$\times \prod_{i=1}^{n} B\left(\frac{\mu_d t_i^{\alpha_i/2}}{\mu_d t_i^{\alpha_i/2} + \mu_c \dfrac{\left(1-\sqrt{\rho_d}\right)}{\left(1-\sqrt{\rho_c}\right)} S_i^{\alpha_i/2}}, \mu_d + k_i, \mu_c + l_i\right), \quad (6.34)$$

with $B(z,a,b)$ being the incomplete beta function [8, Eq. (8.391)].

Finally, the CDF of the multibranch SIR-based SC output can be derived from (6.33), equating the arguments $t_1 = \cdots = t_n = t$:

$$F_\lambda(t) = \underbrace{\sum_{k_1,\ldots,k_n=0}^{\infty}\sum_{l_1,\ldots,l_n=0}^{\infty}}_{2n} G_7$$

$$\times \prod_{i=1}^{n} B\left(\frac{\mu_d t^{\alpha_i/2}}{\mu_d t^{\alpha_i/2} + \mu_c \dfrac{\left(1-\sqrt{\rho_d}\right)}{\left(1-\sqrt{\rho_c}\right)} S_i^{\alpha_i/2}}, \mu_d + k_i, \mu_c + l_i\right). \quad (6.35)$$

Table 6.2 Terms That Need to Be Summed in (6.35) to Achieve Accuracy at the 6th Significant Digit

$S_1/T = 10$ dB, $\alpha_1 = \alpha_2 = 2$ DUAL-BRANCH SELECTION COMBINING DIVERSITY CASE		$\mu_d = 1$ $\mu_c = 1$	$\mu_d = 1.2$ $\mu_c = 1.5$
$\rho_d = 0.3$	$\rho_c = 0.2$	24	21
$\rho_d = 0.3$	$\rho_c = 0.3$	28	25
$\rho_d = 0.3$	$\rho_c = 0.4$	37	35
$\rho_d = 0.4$	$\rho_c = 0.3$	31	27
$\rho_d = 0.5$	$\rho_c = 0.5$	51	47
$S_1/T = 10$ dB, $\alpha_1 = \alpha_2 = \alpha_3 = 2$ TRIPLE-BRANCH SELECTION COMBINING DIVERSITY CASE		$\mu_d = 1$ $\mu_c = 1$	$\mu_d = 1.2$ $\mu_c = 1.5$
$\rho_d = 0.3$	$\rho_c = 0.2$	11	11
$\rho_d = 0.3$	$\rho_c = 0.3$	17	16
$\rho_d = 0.3$	$\rho_c = 0.4$	23	22
$\rho_d = 0.4$	$\rho_c = 0.3$	20	19

Note: We consider a dual- and triple-branch selection combining diversity system.

The nested n infinite sum in (6.35) converges for any value of the parameters ρ_c, ρ_d, μ_d, μ_c, and S_i, which is shown in Table 6.2. The terms needed to be summed to achieve a desired accuracy depend strongly on the correlation coefficients, ρ_d and ρ_c, $0 \leq \rho_d < 1$, $0 \leq \rho_c < 1$. It is obvious that the number of the terms increases as correlation coefficients increase. As shown in Table 6.2, for a higher number of diversity branches, convergence to an acceptable accuracy is slower and we need much more terms.

The PDF at the output of the SC can be obtained easily from the previous expression:

$$f_\lambda(t) = \frac{d}{dt} F_\lambda(t) = \sum_{k_1=0}^{\infty} \cdots \sum_{k_n} \sum_{l_1}^{\infty} \cdots \sum_{l_n}^{\infty} \frac{G_7}{2t} \times \sum_{i=1}^{n} A_i(t)$$

$$A_i(t) = \alpha_i \left(\frac{\mu_d t^{\alpha_i/2}}{\mu_c \frac{\left(1 - \sqrt{\rho_d}\right)}{\left(1 - \sqrt{\rho_c}\right)} S_i^{\alpha_i/2} + \mu_d t^{\alpha_i/2}} \right)^{\mu_d + k_i}$$

$$\times \left(\frac{\mu_c \dfrac{\left(1-\sqrt{\rho_d}\right)}{\left(1-\sqrt{\rho_c}\right)} S_i^{\alpha_i/2}}{\mu_c \dfrac{\left(1-\sqrt{\rho_d}\right)}{\left(1-\sqrt{\rho_c}\right)} S_i^{\alpha_i/2} + \mu_d t^{\alpha_i/2}} \right)^{\mu_c + l_i}$$

$$\times \prod_{\substack{j=1,\dots,n \\ j \neq i}} B\left(\frac{\mu_d t^{\alpha_j/2}}{\mu_c \dfrac{\left(1-\sqrt{\rho_d}\right)}{\left(1-\sqrt{\rho_c}\right)} S_j^{\alpha_j/2} + \mu_d t^{\alpha_j/2}}, \mu_d + k_j, \mu_c + l_j \right).$$

$$(6.36)$$

One can see in Figure 6.13 the PDF of a triple-branch SC output SIR for balanced and unbalanced SIRs at the input branches and some values of correlation coefficients and fading severity parameters.

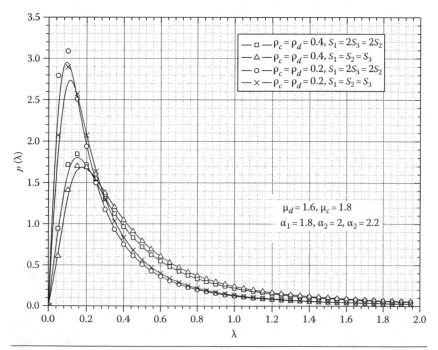

Figure 6.13   PDF of triple-branch SC output SIR for various values of correlation coefficients and fading severity parameters.

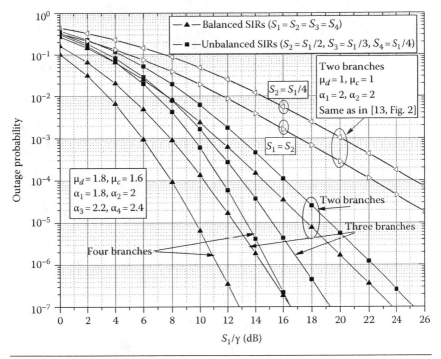

Figure 6.14 Outage probability versus normalized parameter S_1/γ.

In Figure 6.14, the OP is plotted versus normalized parameter S_1 by the protection ratio γ_{th}. The OP is depicted for equal and unequal branch input mean SIRs and diversity systems with two, three, and four branches. Figure 6.14 illustrates that OP could be considerably degraded due to branches unbalancing. Also, from the results in this figure, one can notice the influence of diversity order on OP. At OP of 10^{-6}, in the case of balancing branches, when n increases from $n = 2$ to $n = 3$, that is, from $n = 3$ to $n = 4$, the corresponding additional diversity gains are 6.2 and 9.5 dB, respectively. For the special case of $n = 2$, $m_d = m_c = 1$, $\alpha_1 = \alpha_1 = 2$, same results are obtained as those from [10, Eq. (7)]. The corresponding two dependences for $S_1 = S_2$ and $S_2 = S_1/4$ are presented in this figure, and they are the same as the corresponding two curves in [10, Fig. 2].

Figure 6.15 illustrates the influence of correlation coefficients ρ_d and ρ_c on OP. It is shown that OP increases when correlation among branches increases. At OP of 10^{-5}, if the correlation coefficient increases from 0.1 to 0.4, the corresponding penalties in S_1/γ are from 0.8 dB to 1.5 dB. Also, the effect of fading

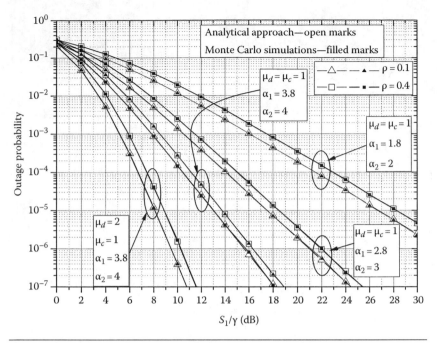

Figure 6.15   Influence of correlation coefficients ρ_d and ρ_c on outage probability.

severity can be noticed in this figure. In the case of $\rho_d = \rho_c = 0.1$, for $S_1/\gamma = 10$ dB, OP increases about 400 times if μ_d increases from 1 to 2. In this figure, the numerical results obtained by using (6.35) are confirmed by Monte Carlo simulations. The simulation results are obtained on the basis of 10^9 samples generated by using C++ programming language. By using all proposed relationships between Nakagami-m and α-μ variables (as well as their corresponding correlation coefficients) and the algorithm for generating two correlated Nakagami-m variables from [17], we generate samples of correlated α-μ variables. For generating gamma variables, which are necessary in the process of generating Nakagami-m variables, we use the algorithm from [18].

In Figure 6.16, the OP for triple-branch SC receiver is plotted versus correlation coefficient ρ_d, for several values of S_1/γ, μ_d, and μ_c. Also balanced ($S_1 = S_2 = S_3$) and unbalanced ($S_2 = S_1/2$, $S_3 = S_1/3$) SIRs at the input branches are observed. It is obvious that for $S_1/\gamma = 0$ dB (strong interference) the OP increases slowly as the correlation coefficient increases, while an increase in μ_d, μ_c does not have a significant effect

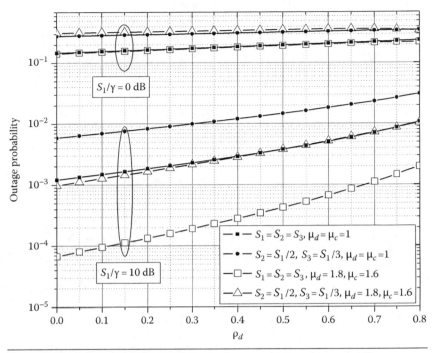

Figure 6.16 Influence of fading severity parameters μ_d and μ_c on outage probability.

on OP. For higher values of S_1/γ, as in the case of $S_1/\gamma = 10$ dB, the influence of μ_d, μ_c, and ρ_d on OP becomes stronger.

In Figure 6.17 is shown how ASEP changes versus S_1(dB) for $\mu_d = 1.2$ and $\mu_c = 1.6$ and various values of correlation coefficients ρ_d and ρ_c, observing the BDPSK modulation scheme. A diversity system with two branches is assumed. Balanced ($S_1 = S_2$) and unbalanced ($S_2 = S_1/2$) SIRs at the input are analyzed. In the same figure, ASEP is depicted for BFSK, assuming the same values for system parameters as for BDPSK. It is obvious that as signal and interference correlation coefficients ρ_d and ρ_c decrease as S_1(dB) increases, the average SEP decreases (which means better system performance). Also, from Figure 6.17, it is visible how the BDPSK modulation scheme outperforms the BFSK modulation scheme. The results obtained by performing a numerical integration using Equation 6.35 are presented, along with the results obtained by Monte Carlo integration [19]. The Monte Carlo integration is performed by using 10^9 samples. The good match between these results is evident.

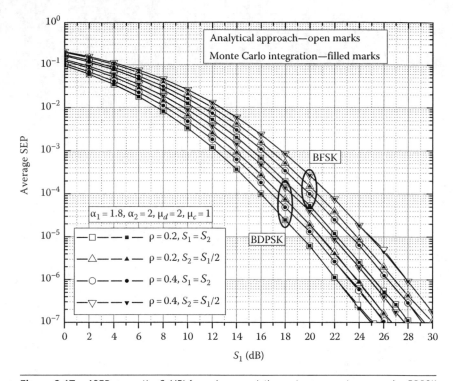

Figure 6.17   ASEP versus the S_1 (dB) for various correlation system parameters assuming BDPSK modulation scheme.

6.1.5 SC Diversity Reception over Exponentially Correlated α-μ Fading Channels

Let us consider the α-μ model of distribution with exponential correlation. We are assuming arbitrary correlation coefficients between fading signals and between interferences, because correlation coefficients depend on the arrival angles of the contribution with the broadside directions of antennas, which are in general case arbitrary [19]. The exponential correlation model [20] can be obtained by setting $\Sigma_{i,j} \equiv \rho^{|i-j|}$ for in correlation matrix, for both the desired signal and interference, where ρ denotes the power correlation coefficient defined as $cov(R_i^2, R_j^2)/(var(R_i^2)var(R_j^2))^{1/2}$.

There is a need to derive the joint statistics for multi α-μ variables. We are relying on some results that are already available in the literature for the exponential correlation model of the Nakagami-*m* distribution [20, Eq. (3)]. After using the same transformation already explained in the previous section, joint distributions of the PDF for both the desired and interfering signal correlated envelopes for multi-branch signal combiner could be expressed as

$$f_{R_1,R_2,\ldots,R_n}(R_1,R_2,\ldots,R_n)$$

$$= \sum_{k_1,k_2,\ldots,k_{n-1}=0}^{\infty} \alpha_1\alpha_2\cdots\alpha_n \frac{\rho_d^{2(k_1+k_2+\cdots+k_{n-1})}}{2^{n\mu_d+2(k_1+k_2+\cdots+k_{n-1})}\Gamma(\mu_d)}$$

$$\times \Gamma(\mu_d+k_1)\ldots\Gamma(\mu_d+k_{n-1})k_1!\ldots k_{n-1}!$$

$$\times \frac{1}{\left(1-\rho_d^2\right)^{(n-1)\mu_d+2(k_1+k_2+\ldots+k_{n-1})}} \times \exp\left(-\frac{R_1^{\alpha_1}+R_n^{\alpha_n}}{2(1-\rho_d^2)}-G_8\right)$$

$$\times R_1^{\alpha_1(\mu_d+k_1)-1} R_n^{\alpha_n(\mu_d+k_{n-1})-1} \times G_9;$$

$$G_8 = \begin{cases} 0, n=2 \\ \dfrac{(\rho_d^2+1)}{2(1-\rho_d^2)}\displaystyle\sum_{i=2}^{n-1} R_i^{\alpha_i}, n>2 \end{cases}; \quad G_9 = \begin{cases} 1, n=2 \\ \displaystyle\prod_{i=2}^{n-1} R_i^{\alpha_i(\mu_d+k_{i-1}+k_i)-1}, n>2 \end{cases};$$

$$(6.37)$$

$$f_{r_1,r_2,\ldots,r_n}(r_1,r_2,\ldots,r_n)$$

$$= \sum_{l_1,l_2,\ldots,l_{n-1}=0}^{\infty} \alpha_1\alpha_2\cdots\alpha_n \frac{\rho_c^{2(l_1+l_2+\cdots+l_{n-1})}}{2^{n\mu_c+2(l_1+l_2+\cdots+l_{n-1})}\Gamma(\mu_c)}$$

$$\times \Gamma(\mu_c+l_1)\ldots\Gamma(\mu_c+l_{n-1})l_1!\ldots l_{n-1}!$$

$$\times \frac{1}{\left(1-\rho_c^2\right)^{(n-1)\mu_c+2(l_1+l_2+\cdots+l_{n-1})}} \times \exp\left(-\frac{r_1^{\alpha_1}+r_n^{\alpha_n}}{2(1-\rho_c^2)}-G_{10}\right)$$

$$\times r_1^{\alpha_1(\mu_c+l_1)-1} r_n^{\alpha_n(\mu_c+l_{n-1})-1} \times G_{11};$$

$$G_{10} = \begin{cases} 0, n=2 \\ \dfrac{(\rho_c^2+1)}{2(1-\rho_c^2)}\displaystyle\sum_{i=2}^{n-1} r_i^{\alpha_i}, n>2 \end{cases}; \quad G_{11} = \begin{cases} 1, n=2 \\ \displaystyle\prod_{i=2}^{n-1} r_i^{\alpha_i(\mu_c+l_{i-1}+l_i)-1}, n>2 \end{cases},$$

$$(6.38)$$

where, ρ_d and ρ_c are the correlation coefficients and μ_d and μ_c are fading severity parameters for the desired and interference signals, correspondingly. Instantaneous values of SIR at the kth diversity branch input can be defined as $\lambda_k = R_k^2/r_k^2$. The joint PDF of instantaneous values of SIRs at the n input branches of SC, λ_k, $k = 1, 2, \ldots, n$, can be given according to Equation 6.30.

After substituting (6.37) and (6.38) in (6.30), the joint PDF can be written as

$$
f_{\lambda_1,\lambda_2,\ldots,\lambda n}\left(t_1,t_2,\ldots,t_n\right) = \sum_{k_1,k_2,\ldots,k_{n-1}=0}^{\infty} \sum_{l_1,l_2,\ldots,l_{n-1}=0}^{\infty} \frac{\alpha_1\alpha_2\ldots\alpha_n}{2^n}
$$

$$
\times \frac{\Gamma\left(\mu_d+\mu_c+k_1+l_1\right)\Gamma\left(\mu_d+\mu_c+k_{n-1}+l_{n-1}\right)}{\Gamma\left(\mu_d\right)\Gamma\left(\mu_c\right)\prod_{i=1}^{n-1}\Gamma\left(\mu_d+k_i\right)\Gamma\left(\mu_c+l_i\right)k_i!\,l_i!} \times G_{12}
$$

$$
\times \rho_d^{2(k_1+k_2+\cdots+k_{n-1})}\rho_c^{2(l_1+l_2+\cdots+l_{n-1})}\left(1-\rho_d^2\right)^{n\mu_c+\mu_d+2(l_1+l_2+\cdots+l_{n-1})}
$$

$$
\times\left(1-\rho_c^2\right)^{n\mu_d+\mu_c+2(k_1+k_2+\cdots+k_{n-1})} \times \frac{t_1^{(\alpha_1(\mu_d+k_1)/2)-1}}{\left(\left(1-\rho_c^2\right)t_1^{\frac{\alpha_1}{2}}+\left(1-\rho_d^2\right)\right)^{\mu_d+\mu_c+k_1+l_1}}
$$

$$
\times \frac{t_n^{\alpha_n(\mu_d+k_{n-1})/2-1}}{\left(\left(1-\rho_c^2\right)t_n^{\alpha_n/2}+\left(1-\rho_d^2\right)\right)^{\mu_d+\mu_c+k_{n-1}+l_{n-1}}} \times G_{13}
$$

$$
G_{12} = \left\{ \begin{array}{l} 1, n = 2 \\ \displaystyle\prod_{i=2}^{n-1}\Gamma\left(\mu_d+\mu_c+k_{i-1}+l_{i-1}+k_i+l_i\right), n > 2 \end{array} \right\};
$$

$$
G_{13} = \left\{ \begin{array}{l} 1, n = 2 \\ \displaystyle\prod_{i=2}^{n-1}\frac{t_i^{\alpha_i(\mu_d+k_{i-1}+k_i)/2-1}}{\left(\left(1-\rho_c^2\right)\left(1+\rho_d^2\right)t_i^{\alpha_i/2}\right.} \\ \left. +\left(1-\rho_d^2\right)\left(1+\rho_c^2\right)\right)^{\mu_d+\mu_c+k_{i-1}+l_{i-1}+k_i+l_i}, n > 2 \end{array} \right\}. \quad (6.39)
$$

Similarly, the joint CDF can be written by substituting expression (6.39) in (6.33), and after n successive integrations, the joint CDF becomes

$$
F_\lambda(t) = \sum_{k_1,k_2,\ldots,k_{n-1}=0}^{\infty} \sum_{l_1,l_2,\ldots,l_{n-1}=0}^{\infty} \frac{\begin{aligned}\Gamma(\mu_d + \mu_c + k_1 + l_1)\\ \times\Gamma(\mu_d + \mu_c + k_{n-1} + l_{n-1})\end{aligned}}{\Gamma(\mu_d)\Gamma(\mu_c)\prod_{i=1}^{n-1}\Gamma(\mu_d + k_i)}
$$

$$
\times\Gamma(\mu_c + l_i)k_i!l_i!
$$

$$
\times G_{10}\frac{\rho_d^{2(k_1+k_2+\cdots+k_{n-1})}\rho_c^{2(l_1+l_2+\cdots+l_{n-1})}\left(1-\rho_d^2\right)^{\mu_d}\left(1-\rho_c^2\right)^{\mu_c}}{\left(1+\rho_d^2\right)^{\mu_d+k_1+k_2+\cdots+k_{n-1}}\left(1+\rho_c^2\right)^{\mu_c+l_1+l_2+\cdots+l_{n-1}}}
$$

$$
\times B\left(\frac{t_1^{\alpha_1/2}}{\dfrac{\left(1-\rho_d^2\right)}{\left(1-\rho_c^2\right)}+t_1^{\alpha_1/2}},\mu_d + k_1,\mu_c+l_1\right)
$$

$$
\times B\left(\frac{t_n^{\alpha_n/2}}{\dfrac{\left(1-\rho_d^2\right)}{\left(1-\rho_c^2\right)}+t_n^{\alpha_n/2}},\mu_d + k_{n-1},\mu_c+l_{n-1}\right)\times G_{12};
$$

$$
G_{12} = \begin{cases}1, n=2\\[2ex]\displaystyle\prod_{i=2}^{n-1}B\left(\frac{t_i^{\alpha_i/2}}{\dfrac{\left(1-\rho_d^2\right)\left(1+\rho_c^2\right)}{\left(1-\rho_c^2\right)\left(1+\rho_d^2\right)}+t_i^{\alpha_i/2}},\right.\\[4ex]\qquad\left.\times\mu_d + k_{i-1} + k_i,\mu_c+l_{i-1} + l_i\right), n>2\end{cases}. \tag{6.40}
$$

The CDF of the multibranch SIR-based SC output can be derived from (6.40) by equating the arguments $t_1 = t_2 = \cdots = t_n = t$:

$$F_\lambda(t) = \sum_{k_1,k_2,\ldots,k_{n-1}=0}^{\infty} \sum_{l_1,l_2,\ldots,l_{n-1}=0}^{\infty} \frac{\begin{array}{c}\Gamma\left(\mu_d + \mu_c + k_1 + l_1\right)\\ \times \Gamma\left(\mu_d + \mu_c + k_{n-1} + l_{n-1}\right)\end{array}}{\Gamma\left(\mu_d\right)\Gamma\left(\mu_c\right)}$$

$$\times \prod_{i=1}^{n-1} \Gamma\left(\mu_d + k_i\right)\Gamma\left(\mu_c + l_i\right) k_i! l_i!$$

$$\times G_{12} \frac{\rho_d^{2\left(k_1+k_2+\cdots+k_{n-1}\right)} \rho_c^{2\left(l_1+l_2+\cdots+l_{n-1}\right)} \left(1-\rho_d^2\right)^{\mu_d} \left(1-\rho_c^2\right)^{\mu_c}}{\left(1+\rho_d^2\right)^{\mu_d+k_1+k_2+\cdots+k_{n-1}} \left(1+\rho_c^2\right)^{\mu_c+l_1+l_2+\cdots+l_{n-1}}}$$

$$\times B\left(\frac{t^{\alpha_1/2}}{\dfrac{\left(1-\rho_d^2\right)}{\left(1-\rho_c^2\right)} + t^{\alpha_1/2}}, \mu_d+k_1, \mu_c+l_1\right)$$

$$\times B\left(\frac{t^{\alpha_n/2}}{\dfrac{\left(1-\rho_d^2\right)}{\left(1-\rho_c^2\right)} + t^{\alpha_n/2}}, \mu_d+k_{n-1}, \mu_c+l_{n-1}\right) \times G_{15};$$

$$G_{15} = \begin{cases} 1, n=2 \\ \displaystyle\prod_{i=2}^{n-1} B\left(\dfrac{t^{\alpha_i/2}}{\dfrac{\left(1-\rho_d^2\right)\left(1+\rho_c^2\right)}{\left(1-\rho_c^2\right)\left(1+\rho_d^2\right)} + t^{\alpha_i/2}}, \\ \qquad\qquad \mu_d + k_{i-1} + k_i, \mu_c + l_{i-1} + l_i\right), n > 2 \end{cases}. \qquad (6.41)$$

The PDF at the output of the SC can be obtained easily from the previous expression:

$$f_\lambda(t) = \frac{d}{dt} F_\lambda(t) = \sum_{k_1=0}^{\infty} \sum_{k_2=0}^{\infty} \cdots \sum_{k_{n-1}=0}^{\infty} \sum_{l_1=0}^{\infty} \sum_{l_2=0}^{\infty} \cdots \sum_{l_{n-1}=0}^{\infty} \frac{1}{2t}$$

$$\times \frac{\Gamma(\mu_d + \mu_c + k_1 + l_1)\Gamma(\mu_d + \mu_c + k_{n-1} + l_{n-1})}{\Gamma(\mu_d)\Gamma(\mu_c)\prod_{i=1}^{n-1}\Gamma(\mu_d + k_i)\Gamma(\mu_c + l_i)k_i!l_i!} \times G_{12}$$

$$\times \frac{\rho_d^{2(k_1+k_2+\cdots+k_{n-1})}\rho_c^{2(l_1+l_2+\cdots+l_{n-1})}\left(1-\rho_d^2\right)^{\mu_d}\left(1-\rho_c^2\right)^{\mu_c}}{\left(1+\rho_d^2\right)^{\mu_d+k_1+k_2+\cdots+k_{n-1}}\left(1+\rho_c^2\right)^{\mu_c+l_1+l_2+\cdots+l_{n-1}}}$$

$$\times \left(A_1(t) + \sum_{i=2}^{n-1} A_i(t) + A_n(t) \right), \tag{6.42}$$

with

$$A_1(t) = \alpha_1 \left(\frac{1}{1 + \frac{\left(1-\rho_c^2\right)}{\left(1-\rho_d^2\right)} t^{\alpha_1/2}} \right)^{\mu_c+l_1} \left(\frac{t^{\alpha_1/2}}{\frac{\left(1-\rho_d^2\right)}{\left(1-\rho_c^2\right)} + t^{\alpha_1/2}} \right)^{\mu_d+k_1}$$

$$\times B\left(\frac{t^{\alpha_2/2}}{\frac{\left(1-\rho_d^2\right)\left(1+\rho_c^2\right)}{\left(1-\rho_c^2\right)\left(1+\rho_d^2\right)} + t^{\alpha_2/2}}, \mu_d + k_1 + k_2, \mu_c + l_1 + l_2 \right) \times \cdots$$

$$\times B\left(\frac{t^{\alpha_n/2}}{\frac{\left(1-\rho_d^2\right)}{\left(1-\rho_c^2\right)} + t^{\alpha_n/2}}, \mu_d + k_n, \mu_c + l_n \right);$$

$$A_i(t) = \alpha_i \left(\frac{1}{1 + \dfrac{\left(1 - \rho_c^2\right)\left(1 + \rho_d^2\right)}{\left(1 - \rho_d^2\right)\left(1 + \rho_c^2\right)} t^{\alpha_i/2}} \right)^{\mu_c + l_{i-2} + l_{i-1}}$$

$$\times \left(\frac{t^{\alpha_i/2}}{\dfrac{\left(1 - \rho_d^2\right)\left(1 + \rho_c^2\right)}{\left(1 - \rho_c^2\right)\left(1 + \rho_d^2\right)} + t^{\alpha_i/2}} \right)^{\mu_d + k_{i-2} + k_{i-1}}$$

$$\times B \left(\frac{t^{\alpha_1/2}}{\dfrac{\left(1 - \rho_d^2\right)}{\left(1 - \rho_c^2\right)} + t^{\alpha_1/2}}, \mu_d + k_1, \mu_c + l_1 \right)$$

$$\times B \left(\frac{t^{\alpha_n/2}}{\dfrac{\left(1 - \rho_d^2\right)}{\left(1 - \rho_c^2\right)} + t^{\alpha_n/2}}, \mu_d + k_n, \mu_c + l_n \right)$$

$$\times \prod_{\substack{j=2,\ldots,n-1 \\ j \neq i}} B \left(\frac{t^{\alpha_i/2}}{\dfrac{\left(1 - \rho_d^2\right)\left(1 + \rho_c^2\right)}{\left(1 - \rho_c^2\right)\left(1 + \rho_d^2\right)} + t^{\alpha_i/2}}, \mu_d + k_{i-2} + k_{i-1}, \mu_c + l_{i-2} + l_{i-1} \right);$$

$$A_n(t) = \alpha_n \left(\cfrac{1}{1 + \cfrac{\left(1-\rho_c^2\right)}{\left(1-\rho_d^2\right)} t^{\alpha_n/2}} \right)^{\mu_c+l_{n-1}} \left(\cfrac{t^{\alpha_n/2}}{\cfrac{\left(1-\rho_d^2\right)}{\left(1-\rho_c^2\right)} + t^{\alpha_n/2}} \right)^{\mu_d+k_{n-1}}$$

$$\times B\left(\cfrac{t^{\alpha_1/2}}{\cfrac{\left(1-\rho_d^2\right)}{\left(1-\rho_c^2\right)} + t^{\alpha_1/2}}, \mu_d+k_1, \mu_c+l_1 \right)$$

$$\times B\left(\cfrac{t^{\alpha_2/2}}{\cfrac{\left(1-\rho_d^2\right)\left(1+\rho_c^2\right)}{\left(1-\rho_c^2\right)\left(1+\rho_d^2\right)} + t^{\alpha_2/2}}, \mu_d+k_1+k_2, \mu_c+l_1+l_2 \right) \times \cdots \times$$

$$\times B\left(\cfrac{t^{\alpha_{n-1}/2}}{\cfrac{\left(1-\rho_d^2\right)\left(1+\rho_c^2\right)}{\left(1-\rho_c^2\right)\left(1+\rho_d^2\right)} + t^{\alpha_{n-1}/2}}, \mu_d+k_{n-2}+k_{n-1}, \mu_c+l_{n-2}+l_{n-1} \right). \tag{6.43}$$

Figure 6.18 shows the output SIR PDF of a triple-branch SC receiver for various values of correlation coefficient and fading severity parameters.

In Figure 6.19, the OP is plotted versus $1/\gamma$ for several values of μ_d and μ_c, and correlation coefficients ρ_d and ρ_c. Diversity systems with two and three branches are observed. It is evident that the system with three branches has better performance (lower values of OP for the same parameters). Also it is interesting to note here that for low values of $1/\gamma$ (<2 dB) due to interference dominance, the OP increases

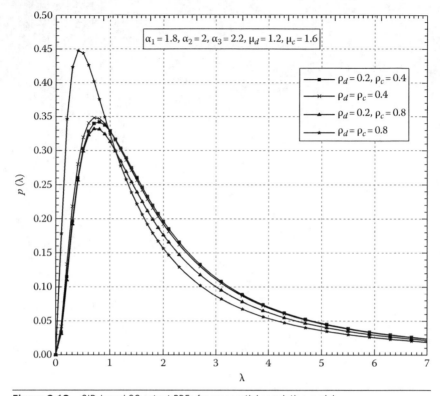

Figure 6.18   SIR-based SC output PDFs for exponential correlation model.

(performance deteriorates) when the fading severity of the interferers decreases (μ_c increases). However, for higher values of $1/\gamma$ (dominance of the desired signal), the OP decreases when μ_c increases.

In Figure 6.20, the OP is plotted versus the correlation coefficient ρ_d for several values of μ_d and μ_c. For $1/\gamma = 0$ dB, the OP increases slowly as the correlation coefficient increases (due to strong interference). In addition, μ_d and ρ_d do not have a significant effect on OP. The influence of μ_d and ρ_d on the OP becomes stronger for higher values of $1/\gamma$ ($1/\gamma = 10$ dB).

The effect of the fading correlation and CCI on the output SIR has been investigated in Figures 6.21 and 6.22. In Figure 6.21, the average output SIR, versus the correlation coefficient ρ_d, is depicted for several values of the fading severities μ_d and μ_c and several values of interference correlation coefficient ρ_c. It is observed here that, with μ_c constant, the effect of ρ_c on the system performance is small due to the inferiority of the correlation between the desired signals.

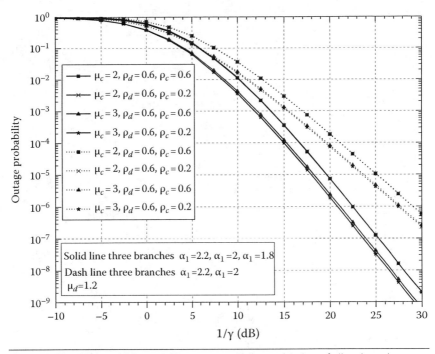

Figure 6.19 SIR-based SC output OP over exponentially correlated α-μ fading channels.

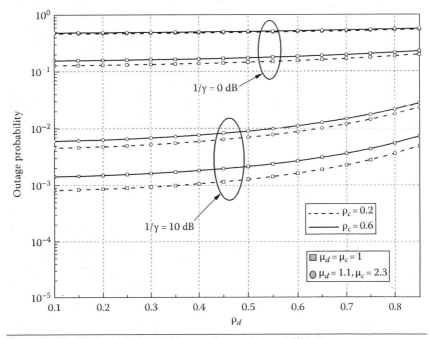

Figure 6.20 SIR-based SC output OP versus the correlation coefficient ρ_d.

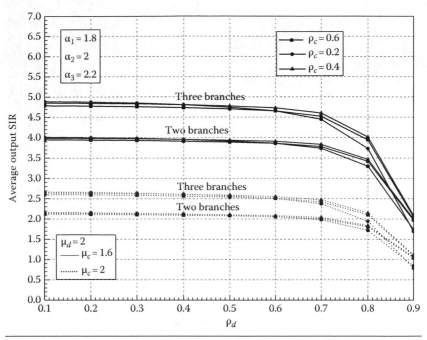

Figure 6.21   Average output SIR versus the power correlation coefficient ρ_d.

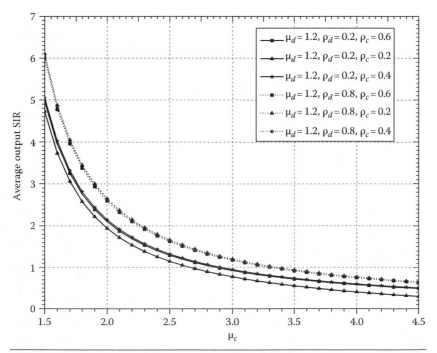

Figure 6.22   Average output SIR versus cochannel fading parameter μ_c.

Moreover, diversity gain decreases (smaller values of average output SIR) with an increase in power correlation coefficient ρ_d, as expected. Also, it is evident that the triple SC system has better performance than the dual one for the same parameters.

Furthermore, it is ascertained that the output SIR decreases as the parameter μ_c increases. In addition, the proposed analysis confirms that the diversity gain changes rapidly with a small change of μ_c. This is shown in Figure 6.22, where the normalized output SIR versus the CCI severity parameter μ_c is plotted. A significant improvement at the diversity gain is observed for lower values of μ_c ($\mu_c < 2$). It is because of the deep fading behavior of the CCIs (lower values of μ_c), which lead to an increase in the average SIR at the output of the SC. It is obvious here that the most effective parameter for the output SIR performance is the fading severity of the CCI signal.

In Figure 6.23, numerical evaluated results for the AoF of dual- and triple-SC receivers in the function of correlation coefficient ρ_d are demonstrated. It is evident that an increase in ρ_d and/or decrease in α fading parameter leads to a corresponding increase in the AoF resulting in performance degradation.

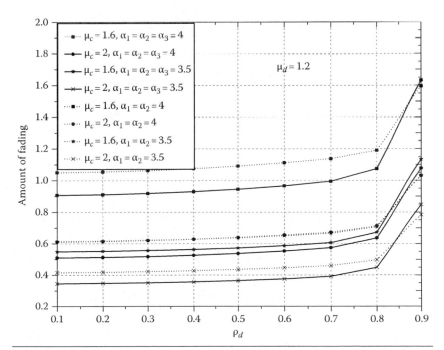

Figure 6.23 AoF of dual- and triple-branch SC receiver versus the correlation coefficient ρ_d.

6.1.6 SC Diversity Reception over Generally Correlated α-μ Fading Channels

The performances of the dual-branch SC reception over generally correlated α-μ fading channels in the presence of CCI can be carried out by considering the joint PDF distribution given in [21]. By using (6.8), we can obtain an expression for the joint CDF and further for the CDF of output SIR, by equating the arguments $t_1 = t_2 = t$ in the form of

$$F_\lambda(t) = \sum_{l=0}^{\infty}\sum_{k=0}^{\infty}\sum_{j=0}^{l}\sum_{m=0}^{l}\sum_{n=0}^{k}\sum_{p=0}^{k} G_{16}G_{17}{}^{\mu_d+j}G_{18}{}^{\mu_d+m}$$

$$\times {}_2F_1\left(\mu_d+j,1-\mu_c-n;1+\mu_1+j;G_{17}\right)$$

$$\times {}_2F_1\left(\mu_d+m,1-\mu_c-p;1+\mu_d+m;G_{18}\right);$$

$$G_{16} = \frac{\begin{array}{c}(-1)^{j+m+n+p}\, l!k!\rho_{12d}^l\Gamma\left(\mu_d+\mu_c+n+j\right)\\ \times\Gamma\left(\mu_d+\mu_c+p+m\right)\end{array}}{\left(\mu_d+j\right)\left(\mu_d+m\right)j!n!m!\,p!\left(l-j\right)!\left(l-m\right)!}$$
$$\Gamma\left(\mu_d\right)\Gamma\left(\mu_c\right)\Gamma\left(\mu_d+l\right)\Gamma\left(\mu_c+k\right)$$

$$\times \frac{\begin{array}{c}\mu_d{}^{2\mu_d+j+p}\mu_c{}^{2\mu_c+n+m}l!k!\\ \times\rho_{12c}^k\left(\left(\mu_d-1+l\right)!\right)^2\left(\left(\mu_c-1+k\right)!\right)^2\end{array}}{\begin{array}{c}\left(k-n\right)!\left(k-p\right)!\left(\mu_d-1+j\right)!\\ \times\left(\mu_d-1+m\right)!\left(\mu_c-1+n\right)!\left(\mu_c-1+p\right)!\end{array}};$$

$$G_{17} = \left(\frac{\mu_d t^{\alpha_1/2}}{\mu_c S_1{}^{\alpha_1/2}+\mu_d t^{\alpha_1/2}}\right);\quad G_{18} = \left(\frac{\mu_d t^{\alpha_2/2}}{\mu_c S_2{}^{\alpha_2/2}+\mu_d t^{\alpha_2/2}}\right), \qquad (6.44)$$

with $S_k = \hat{R}_k^2/\hat{r}_k^2$ being the average SIRs at the kth input branch of the multibranch SC, $k=1,2$. The nested infinite sum in (6.44), for two branches diversity case, converges for any value of the parameters ρ_{d12}, ρ_{d12}, S_1, S_2, μ_d, μ_c, α_1, and α_2. As shown in Table 6.3, the number of the terms needed to be summed to achieve a desired accuracy depends strongly on the correlation coefficients ρ_{d12}, ρ_{d12}. The number of the terms increases as the correlation coefficient increases. For the special case of $\mu_d=1$ and $\mu_c=1$, we can evaluate the expression for CDF for Weibull

Table 6.3 Number of Terms That Need to Be Summed in (6.44) to Achieve Accuracy at Significant Digit in Brackets

$\alpha_1 = 2.2$, $\alpha_1 = 2$, $\mu_d = 1.2$ $\mu_c = 1.5$, $S_1 = S_2$		$S/\lambda = -10$ dB (6th)	$S/\lambda = 0$ dB (6th)
$\rho_d = 0.2$	$\rho_c = 0.2$	5	5
$\rho_d = 0.3$	$\rho_c = 0.3$	6	6
$\rho_d = 0.4$	$\rho_c = 0.3$	6	6
	$\rho_c = 0.4$	6	6
	$\rho_c = 0.5$	7	7
	$\rho_c = 0.5$	7	8
	$\rho_c = 0.6$	7	9
$\rho_d = 0.5$	$\rho_c = 0.5$	7	8
$\rho_d = 0.6$	$\rho_c = 0.6$	7	10

desired signal and CCI, and for the special case of $\alpha_1 = 2$ and $\alpha_2 = 2$ we can evaluate the expression for the CDF for the Nakagami-m desired signal and CCI. This generality of the CDF of the output SIR for a number of fading distributions is the main contribution of our work.

The PDF of the output SIR can be obtained easily from the previous expression:

$$p_\lambda(t) = \frac{d}{dt} F_\lambda(t) = \sum_{l=0}^{\infty} \sum_{k=0}^{\infty} \sum_{j=0}^{l} \sum_{m=0}^{l} \sum_{n=0}^{k} \sum_{p=0}^{k} \frac{G_{16}}{2t}$$

$$\times \left[\frac{\alpha_1}{\mu_d + m} \left(\left(\frac{\mu_d t^{\alpha_1/2}}{\mu_c S_1^{\alpha_1/2} + \mu_d t^{\alpha_1/2}} \right)^{2\mu_d + j + m} \left(\frac{\mu_c S_1^{\alpha_1/2}}{\mu_c S_1^{\alpha_1/2} + \mu_d t^{\alpha_1/2}} \right)^{\mu_c + n} \right. \right.$$

$$\left. \times\, _2F_1\left(\mu_d + m, 1 - \mu_c - p; 1 + \mu_d + m; \left(\frac{\mu_d t^{\alpha_2/2}}{\mu_c S_2^{\alpha_2/2} + \mu_d t^{\alpha_2/2}} \right) \right) \right)$$

$$+ \frac{\alpha_2}{\mu_d + j} \left(\left(\frac{\mu_d t^{\alpha_2/2}}{\mu_c S_2^{\alpha_2/2} + \mu_d t^{\alpha_2/2}} \right)^{2\mu_d + j + m} \left(\frac{\mu_c S_2^{\alpha_2/2}}{\mu_c S_2^{\alpha_2/2} + \mu_d t^{\alpha_2/2}} \right)^{\mu_c + p} \right.$$

$$\left. \left. \times\, _2F_1\left(\mu_1 + j, 1 - \mu_2 - n; 1 + \mu_1 + j; \left(\frac{\mu_1 t^{\alpha_1/2}}{\mu_2 S_1^{\alpha_1/2} + \mu_1 t^{\alpha_1/2}} \right) \right) \right) \right].$$

$$G_{19} = (\mu_d + j)(\mu_d + m) G_{16}. \tag{6.45}$$

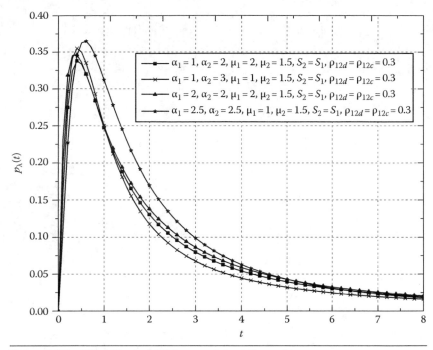

Figure 6.24   PDF of SC output SIR for general correlation model.

Figure 6.24 shows the PDF of the output SIR for balanced input branches and various fading types (various values of α and μ, including special cases of Nakagami-m and Weibull distributions). Comparing values in Figure 6.24 with values of the PDF for the output SIR in the case of SC over correlated Weibull fading channels in the presence of CCI, from [22, Eq. (11)] and Figure 6.24, we can see that for corresponding values of fading parameters, there is a perfect match between these two cases, which is the best validation of our work.

OP versus normalized parameter S_1/γ for balanced and unbalanced ratios of the SIR at the input of the branches and for various values of correlation coefficient and fading severity parameters is shown in Figure 6.25.

6.2 Diversity Reception over Rician Fading Channels in the Presence of CCI

6.2.1 SSC Diversity Reception with Uncorrelated Branches

By using Equations 6.1 and 6.2, where uncorrelated branches at the terminal are considered, and SIR values at antennas, λ_i, are assumed

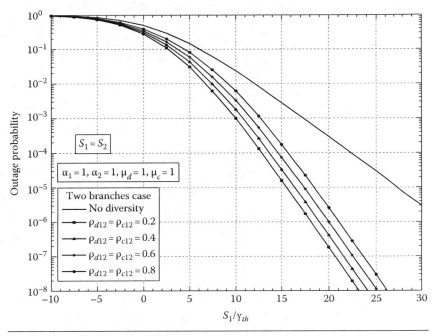

Figure 6.25 OP of SC output SIR for general correlation model.

to be statistically independent, the PDF of the output SIR, λ, can be presented capitalizing on

$$f_{\lambda 1}(\lambda) = \sum_{p=0}^{\infty}\sum_{q=0}^{\infty} \frac{\begin{array}{c}\lambda^p S_1^{q+1} K_{d1}^p K_{c1}^q \left(1+K_{d1}\right)^{p+1}\\ \times \left(1+K_{c1}\right)^{q+1}\Gamma(p+q+2)\\ \hline e^{K_{d1}+K_{c1}}\Gamma\left(p+1\right)p!\Gamma\left(q+1\right)q!\end{array}}{\left(\lambda\left(1+K_{d1}\right)+S_1\left(1+K_{c1}\right)\right)^{p+q+2}};$$

$$f_{\lambda 2}(\lambda) = \sum_{p=0}^{\infty}\sum_{q=0}^{\infty} \frac{\begin{array}{c}\lambda^p S_2^{q+1} K_{d2}^p K_{c2}^q \left(1+K_{d2}\right)^{p+1}\\ \times \left(1+K_{c2}\right)^{q+1}\Gamma(p+q+2)\\ \hline e^{K_{d2}+K_{c2}}\Gamma\left(p+1\right)p!\Gamma\left(q+1\right)q!\end{array}}{\left(\lambda\left(1+K_{d2}\right)+S_1\left(1+K_{c2}\right)\right)^{p+q+2}};$$

$$F_{\lambda_1}(z_T) = \int_0^{z_T} f_{\lambda_1}(\lambda)\,d\lambda = \sum_{p=0}^{\infty} \sum_{q=0}^{\infty} \left[\frac{K_{d1}^p K_{c1}^q \Gamma(p+q+2)}{e^{K_{d1}+K_{c1}} \Gamma(p+1)\,p!\,\Gamma(q+1)q!} \right.$$

$$\times \frac{\left(z_T (1+K_{d1}) / \left(z_T (1+K_{d1}) + S_1 (1+K_{c1}) \right) \right)^{p+1}}{(p+1)}$$

$$\left. \times {}_2F_1\left(p+1, -q; p+2; \frac{z_T (1+K_{d1})}{\left(z_T (1+K_{d1}) + S_1 (1+K_{c1}) \right)} \right) \right];$$

$$F_{\lambda_2}(z_T) = \int_0^{z_T} f_{\lambda_2}(\lambda)\,d\lambda = \sum_{p=0}^{\infty} \sum_{q=0}^{\infty} \left[\frac{K_{d2}^p K_{c2}^q \Gamma(p+q+2)}{e^{K_{d2}+K_{c2}} \Gamma(p+1)\,p!\,\Gamma(q+1)q!} \right.$$

$$\times \frac{\left(\dfrac{z_T (1+K_{d2})}{\left(z_T (1+K_{d2}) + S_2 (1+K_{c2}) \right)} \right)^{p+1}}{(p+1)}$$

$$\left. \times {}_2F_1\left(p+1, -q; p+2; \frac{z_T (1+K_{d2})}{\left(z_T (1+K_{d2}) + S_2 (1+K_{c2}) \right)} \right) \right]. \qquad (6.46)$$

Similarly, the CDF at the output of SIR-based SSC combiner with uncorrelated branches can be determined from Equation 6.5, capitalizing on

$$F_{\lambda_1}(\lambda) = \sum_{p=0}^{\infty} \sum_{q=0}^{\infty} \left[\frac{K_{d1}^p K_{c1}^q \Gamma(p+q+2)}{e^{K_{d1}+K_{c1}} \Gamma(p+1)\,p!\,\Gamma(q+1)q!} \right.$$

$$\times \frac{\left(\lambda (1+K_{d1}) / \left(\lambda (1+K_{d1}) + S_1 (1+K_{c1}) \right) \right)^{p+1}}{(p+1)}$$

$$\left. \times {}_2F_1\left(p+1, -q; p+2; \frac{\lambda (1+K_{d1})}{\left(\lambda (1+K_{d1}) + S_1 (1+K_{c1}) \right)} \right) \right];$$

$$F_{\lambda 2}(\lambda) = \sum_{p=0}^{\infty} \sum_{q=0}^{\infty} \left[\frac{K_{d2}^p K_{c2}^q \Gamma(p+q+2)}{e^{K_{d2}+K_{c2}} \Gamma(p+1) p! \Gamma(q+1) q!} \right.$$

$$\times \frac{\left(\lambda(1+K_{d2}) / \left(\lambda(1+K_{d2}) + S_1(1+K_{c2}) \right) \right)^{p+1}}{(p+1)}$$

$$\times \left. {}_2F_1\left(p+1, -q; p+2; \frac{\lambda(1+K_{d2})}{\left(\lambda(1+K_{d2}) + S_2(1+K_{c2}) \right)} \right) \right]. \quad (6.47)$$

Also based on expressions presented here, standard performance measures analysis can be efficiently carried out.

6.2.2 SSC Diversity Reception with Correlated Branches

Let us consider an SSC diversity receiver applied on a small terminal with insufficient antennae spacing between the antennas. Now, the desired signal envelopes R_1 and R_2 will experience correlative Rician fading, with joint distribution [23]:

$$f_{R_1, R_2}(R_1, R_2) = \frac{R_1 R_2 (K_d + 1)^2}{\beta_d^2 (1 - \rho^2)}$$

$$\times \exp\left(-\frac{(R_1^2 + R_2^2)(K_d + 1) + 4K_d \beta_d (1 - \rho)}{2\beta_d (1 - \rho^2)} \right)$$

$$\times \sum_{k=0}^{\infty} \varepsilon_k I_k \left(\frac{R_1 R_2 \rho (K_d + 1)}{\beta_d (1 - \rho^2)} \right)$$

$$\times I_k \left(\frac{R_1}{(1+\rho)} \sqrt{\frac{2K_d(K_d + 1)}{\beta_d}} \right) I_k \left(\frac{R_2}{(1+\rho)} \sqrt{\frac{2K_d(K_d + 1)}{\beta_d}} \right),$$

$$(6.48)$$

where

β_d denotes the average power of the desired signals, defined as
$\beta_d = \overline{R_1^2}/2 = \overline{R_2^2}/2$

$I_n(x)$ is the nth-order-modified Bessel function of the first kind function [8, Eq. (8.406)]

ρ stands for the correlation coefficient between the desired signals parameter ε_k is defined as $\varepsilon_k = 1$ $(k = 0)$, and $\varepsilon_k = 2$ $(k \neq 0)$

Similarly, the joint PDF of the correlated interference signal envelopes r_1 and r_2 could be expressed by [23]

$$f_{r_1,r_2}(r_1,r_2) = \frac{r_1 r_2 (K_i+1)^2}{\beta_i^2 (1-\rho^2)} \exp\left(-\frac{(r_1^2+r_2^2)(K_i+1)+4K_i\beta_i(1-\rho)}{2\beta_i(1-\rho^2)}\right)$$

$$\times \sum_{l=0}^{\infty} \varepsilon_l I_l\left(\frac{r_1 r_2 \rho (K_i+1)}{\beta_i(1-\rho^2)}\right) I_l\left(\frac{r_1}{(1+\rho)}\sqrt{\frac{2K_i(K_i+1)}{\beta_i}}\right)$$

$$\times I_l\left(\frac{r_2}{(1+\rho)}\sqrt{\frac{2K_i(K_i+1)}{\beta_i}}\right). \tag{6.49}$$

It has been assumed that the correlation level between desired signal envelopes at the reception is the same as the correlation level between the interference signal branches.

Let $\lambda_1 = R_1^2/r_1^2$ and $\lambda_2 = R_2^2/r_2^2$ represent the instantaneous SIR on the diversity branches, respectively.

After substituting (6.48) and (6.49) in (6.8), we obtain the joint PDF in the form of

$$f_{\lambda_1,\lambda_2}(t_1,t_2) = \sum_{k,l,m,n,p,q,s,w=0}^{\infty} \frac{\begin{array}{c} K_d^{n+p+k} K_i^{s+w+l}(K_d+1)^{2m+2k+n+p+1} \\ \times (K_i+1)^{2q+2l+s+w+1} \varepsilon_k \varepsilon_l \rho^{2m+2q+k+l} \\ \times (1-\rho^2)^{n+p+k+s+w+l+2} \end{array}}{\begin{array}{c} \beta_d^{-(2q+2l+s+w+2)} \beta_i^{-(2m++2k+n+p+2)} \\ \times (1+\rho)^{2(n+p+k+s+w+l)} m!n!p!q!s!w! \end{array}}$$

$$\times \frac{t_1^{m+n+k} t_2^{m+p+k}}{[t_1\beta_i(K_d+1)+\beta_d(K_i+1)]^{m+n+k+q+s+l+2}}$$

$$\times [t_2\beta_i(K_d+1)+\beta_d(K_i+1)]^{m+p+k+q+w+l+2}$$

$$\Gamma\left(m+n+k+q+s+l+2\right)$$

$$\times\frac{\times\Gamma\left(m+p+k+q+w+l+2\right)}{\Gamma\left(m+k+1\right)\Gamma\left(n+k+1\right)\Gamma\left(p+k+1\right)}$$

$$\times\Gamma\left(q+l+1\right)\Gamma\left(s+l+1\right)\Gamma\left(w+l+1\right)$$

$$\times\exp\left(-\frac{2\left(K_d+K_i\right)}{1+r}\right). \tag{6.50}$$

Let λ represent the instantaneous SIR at the SSC output, and z_T represent the predetermined switching threshold for both input branches. The PDF of λ is given by (6.11) and (6.12). After some basic mathematical transformations, for this case $v_{SSC}(\lambda)$ can be expressed as infinite series:

$$v_{SSC}(\lambda)=\sum_{k,l,m,n,p,q,s,w=0}^{\infty}G_{20}G_{21}{}^{m+p+k+1}\frac{\lambda^{m+n+k}S^{q+l+s+1}}{\left[\lambda\left(K_d+1\right)+S\left(K_i+1\right)\right]^{m+n+k+q+s+l+2}}$$

$$\times{}_2F_1\left(m+p+k+1,-q-w-l;2+m+p+k;G_{21}\right);$$

$$G_{20}=\varepsilon_k\varepsilon_l\,\frac{\rho^{2m+2q+k+l}\left(1-\rho^2\right)^{n+p+k+s+w+l+2}K_d^{n+p+k}K_i^{s+w+l}}{\times\left(K_d+1\right)^{2m+2k+n+p+1}\left(K_i+1\right)^{2q+2l+s+w+1}}{\Gamma\left(m+k+1\right)\Gamma\left(n+k+1\right)\Gamma\left(p+k+1\right)}$$

$$\times\Gamma\left(q+l+1\right)\Gamma\left(s+l+1\right)\Gamma\left(w+l+1\right)$$

$$\times\exp\left(-\frac{2\left(K_d+K_i\right)}{1+r}\right)$$

$$\times\frac{\Gamma\left(m+n+k+q+s+l+2\right)\Gamma\left(m+p+k+q+w+l+2\right)}{\left(1+\rho\right)^{2\left(n+p+k+s+w+l\right)}m!n!\,p!q!s!w!(m+p+k+1)};$$

$$G_{21}=\frac{\left(K_d+1\right)z_\tau}{\left(K_d+1\right)z_\tau+\left(K_i+1\right)S}, \tag{6.51}$$

where $S = \beta_d/\beta_i$ defines the average SIR at the input branch of the dual-branch SC.

Similarly, assuming that due propagation conditions, $K_{d1} = K_{d2} = K_d$ and $K_{i1} = K_{i2} = K_i$, in the same manner, the $f_\lambda(\lambda)$ can be expressed as

$$f_{\lambda_1}(\lambda) = \sum_{k=0}^{\infty} \sum_{l=0}^{\infty} \frac{\begin{array}{c}(K_d+1)^{k+1}(K_i+1)^{l+1}\lambda^k S^{l+1} \\ \times \Gamma(k+l+2) K_d^k K_i^l\end{array}}{\begin{array}{c}\left(\lambda(K_d+1)+(K_i+1)S\right)^{k+l+2} \\ \times \Gamma(k+1)\Gamma(l+1)k!l!\exp(K_d+K_i)\end{array}}. \tag{6.52}$$

According to (6.16), the CDF of the SSC output SIR can be presented as

$$F_{z_1,z_2}(\lambda, z_\tau) = \sum_{k,l,m,n,p,q,s,w=0}^{\infty} G_{22} G_{23}{}^{m+n+k+1} G_{21}{}^{m+p+k+1}$$

$$\times {}_2F_1\left(m+n+k+1, -q-s-l; m+n+k+3; G_{23}\right)$$

$$\times {}_2F_1\left(m+p+k+1, -q-w-l; m+p+3; G_{21}\right);$$

$$F_z(\lambda) = \sum_{k=0}^{\infty} \sum_{l=0}^{\infty} \frac{\Gamma(k+l+2)}{\Gamma(k+1)\Gamma(l+1)k!l!\exp(K_d+K_i)}$$

$$\times \frac{G_{23}{}^{k+1}}{k+1} {}_2F_1\left(k+1, -l, 2+k, G_{23}\right);$$

$$F_z(z_T) = \sum_{k=0}^{\infty} \sum_{l=0}^{\infty} \frac{\Gamma(k+l+2)}{\Gamma(k+1)\Gamma(l+1)k!l!\exp(K_d+K_i)}$$

$$\times \frac{G_{21}{}^{k+1}}{k+1} {}_2F_1\left(k+1, -l, 2+k, G_{21}\right);$$

$$G_{22} = \varepsilon_k \varepsilon_l \frac{\rho^{2m+2q+k+l}\left(1-\rho^2\right)^{n+p+k+s+w+l+2} K_d^{n+p+k} K_i^{s+w+l}}{\begin{array}{c}\Gamma(m+k+1)\Gamma(n+k+1)\Gamma(p+k+1) \\ \times \Gamma(q+l+1)\Gamma(s+l+1)\Gamma(w+l+1)\end{array}}$$

$$\times \exp\left(-\frac{2(K_d + K_i)}{1+r}\right)$$

$$\times \frac{\Gamma(m+n+k+q+s+l+2)\Gamma(m+p+k+q+w+l+2)}{(1+\rho)^{2(n+p+k+s+w+l)}\,m!\,n!\,p!\,q!\,s!\,w!\,(m+p+k+1)(m+n+k+1)};$$

$$G_{23} = \frac{(K_d+1)\lambda}{(K_d+1)\lambda + (K_i+1)S}. \tag{6.53}$$

The OP is presented in Figure 6.26 as the function of the normalized outage threshold (dB) for several values of parameters ρ, K_d, and K_i. The normalized outage threshold (dB) is defined as the average SIRs at the input branch of the balanced dual-branch switched-and-stay combiner, normalized by the specified threshold value z^*. It can be observed from that figure that the OP deteriorates with a decrease in the Rice factor K_d. In addition, the presented results show branch correlation influence on the OP. That is, when correlation coefficients ρ increase, the OP increases.

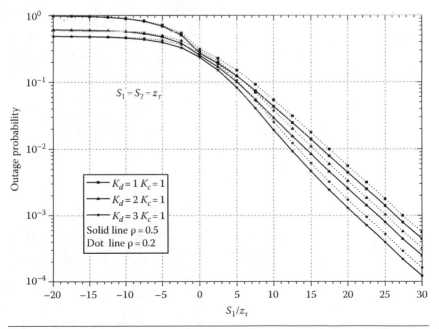

Figure 6.26 OP versus normalized outage threshold for the balanced dual-branch SSC diversity receiver and different values of parameters ρ and K_d.

Using the previously derived infinite series expressions, we can now present representative numerical performance evaluation results of the studied dual-branch SSC diversity receiver, such as ASEP, in the case of two modulation schemes, NCFSK, and BDPSK, for several values of ρ, K_d, and K_i, as a function of the average SIRs at the input branches of the balanced dual-branch SSC, that is, S.

These results are plotted in Figures 6.27 and 6.28. First, a comparison is made between the no-diversity case and the SSC receiving technique. It is obvious that the results are much better when diversity is applied. Then we can observe from the figures that when the average SIR at the input branches and values of the Rice factor of desired signal, K_d, increase, the ASEP increases at the same time. In addition, these figures show better error performances for larger distance between diversity branches, that is, for smaller values of the correlation coefficient ρ. The comparison of Figures 6.27 and 6.28 shows better performance of BDPSK modulation scheme versus NCFSK modulation scheme.

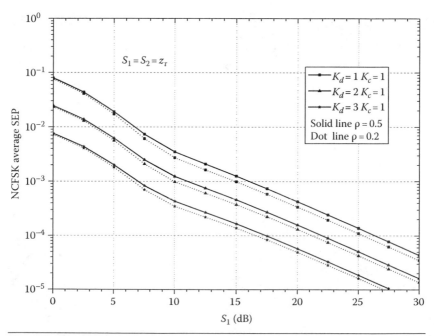

Figure 6.27 ASEP versus average SIRs at the input branches of the balanced SSC, for NCFSK modulation scheme and several values of parameters ρ and K_d.

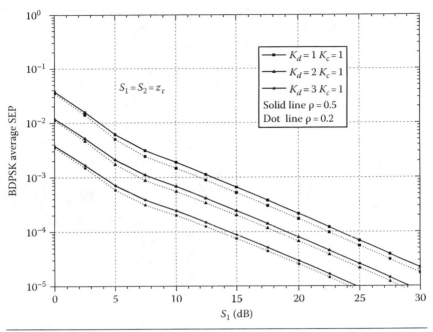

Figure 6.28 BDPSK ASEP versus average SIRs at the input branches of the balanced for several values of parameters ρ and K_d.

6.2.3 SC Diversity Reception with Uncorrelated Branches

For this case, the CDF of the SC output SIR can be determined according to [10] as

$$F_\lambda(t) = F_{\lambda_1}(t)F_{\lambda_2}(t)\cdots F_{\lambda_N}(t) = \prod_{i=1}^{N} F_{\lambda_i}(t)$$

$$
= \prod_{i=1}^{N} \left\{ \sum_{p=0}^{\infty}\sum_{q=0}^{\infty} \left[\frac{K_{di}^{p}K_{ci}^{q}\Gamma(p+q+2)}{e^{K_{di}+K_{ci}}\Gamma(p+1)p!\Gamma(q+1)q!} \right.\right.
$$
$$
\times \frac{\left(t(1+K_{di})/(t(1+K_{di})+S_i(1+K_{ci}))\right)^{p+1}}{(p+1)}
$$
$$
\left.\left. \times {}_2F_1\left(p+1,-q;p+2;\frac{t(1+K_{di})}{(t(1+K_{di})+S_i(1+K_{ci}))}\right) \right]\right\}.
$$

$$(6.54)$$

Then, the PDF of the SC output SIR can be presented as

$$f_\lambda(\lambda) = \frac{d}{d\lambda}F(\lambda) = \sum_{i=1}^{n} f_{\lambda_i}(\lambda) \prod_{\substack{j=1 \\ j \neq i}}^{n} F_{\lambda_i}(\lambda)$$

$$= \sum_{i=1}^{n}\sum_{p=0}^{\infty}\sum_{q=0}^{\infty} \frac{\lambda^p S_i^{q+1} K_{di}^p K_{ci}^q \left(1+K_{di}\right)^{p+1}}{e^{K_{di}+K_{ci}}\Gamma(p+1)p!\Gamma(q+1)}$$

$$\times \frac{\times \left(1+K_{ci}\right)^{q+1}\Gamma(p+q+2)}{\times q!\left(\lambda\left(1+K_{di}\right)+S_1\left(1+K_{ci}\right)\right)^{p+q+2}}$$

$$\times \prod_{\substack{j=1 \\ j \neq i}}^{n}\left\{ \begin{bmatrix} \sum_{r=0}^{\infty}\sum_{t=0}^{\infty}\left[\dfrac{K_{dj}^r K_{cj}^t \Gamma(r+t+2)}{e^{K_{dj}+K_{cj}}\Gamma(r+1)r!\Gamma(t+1)t!} \right. \\[3ex] \times \dfrac{\left(\dfrac{\lambda\left(1+K_{dj}\right)}{\left(\lambda\left(1+K_{dj}\right)+S_j\left(1+K_{cj}\right)\right)}\right)^{r+1}}{(r+1)} \\[3ex] \left. \times\, {}_2F_1\left(r+1,-t;r+2;\dfrac{z_T\left(1+K_{dj}\right)}{\left(\lambda\left(1+K_{dj}\right)+S_j\left(1+K_{cj}\right)\right)}\right) \right] \end{bmatrix} \right\};$$

$$(6.55)$$

with λ_i and S_i denoting instantaneous and average values of the SIR at ith input branch of the SC combiner.

6.2.4 SC Diversity Reception with Correlated Branches

Here we will apply the same approach as in Section 6.1.4, for analyzing performances of a dual-SC SIR-based combiner with correlated branches over Rician fading channels. Introducing expression (6.50) into (6.33), with defining instantaneous values of SIR, at the ith diversity branch as $\lambda_i = R_i^2/r_i^2$, after equating the arguments $\lambda_1 = \lambda_2 = \lambda$, the CDF of the output SIR could be derived in the form of

$$
F_\lambda(t) = \sum_{k,l,m,n,p,q,s,w=0}^{\infty} G_{24} \frac{\lambda^{2m+2k+p+n+2}}{\left(\lambda(K_d+1)+S(K_i+1)\right)^{2m+2k+n+p+2}}
$$

$$
\times {}_2F_1 \left(\begin{array}{c} m+p+k+1,-q-w-l; \\[2mm] \times m+p+k+2; \dfrac{\lambda(K_d+1)}{\lambda(K_d+1)+S(K_i+1)} \end{array} \right)
$$

$$
\times {}_2F_1 \left(\begin{array}{c} m+n+k+1,-q-s-l; \\[2mm] \times m+n+k+2; \dfrac{t(K_d+1)}{\lambda(K_d+1)+S(K_i+1)} \end{array} \right) ;
$$

$$
G_{24} = \frac{(K_d+1)^{2m+2k+p+n+2} \, K_d^{n+p+k} \, K_i^{s+w+l}}{\times \, \varepsilon_k \varepsilon_l \rho^{2m+2q+k+l} \left(1-\rho^2\right)^{n+p+k+s+w+l+2}}{(m+p+k+1)(m+n+k+1)} \exp\left(-\frac{2(K_d+K_i)}{1+\rho}\right)
$$

$$
\times (1+\rho)^{2(n+p+k+s+w+l)} \, m!n!\,p!q!s!w!
$$

$$
\times \frac{\Gamma(m+n+k+q+s+l+2)}{\times\Gamma(m+p+k+q+w+l+2)}{\Gamma(m+k+1)\Gamma(n+k+1)\Gamma(p+k+1)}, \tag{6.56}
$$

$$
\times \Gamma(q+l+1)\Gamma(s+l+1)\Gamma(w+l+1)
$$

where $S = \beta_d/\beta_i$ defines the average SIR at the input branch of the dual-branch SC.

The PDF of the output SIR can be obtained easily from the previous expression:

$$f_\lambda(t) = \frac{d}{dt}F_\lambda(t) = \sum_{k,l,m,n,p,q,s,w=0}^{\infty} G_{25}t^{2m+2k+p+n+1}$$

$$\left[\frac{S^{q+l+t+1}(1+K_i)^t \ {}_2F_1\left(\begin{array}{c} m+n+k+1, -l-q-s; \\ \times m+n+k+2; \\ \times t(K_d+1)/(t(K_d+1)+S(K_i+1)) \end{array}\right)}{\left(t(K_d+1)+S(K_i+1)\right)^{2m+2k+l+n+p+w+q+3}(m+n+k+1)}\right.$$

$$\left.+\frac{S^{q+l+s+1}(1+K_i)^s \ {}_2F_1\left(\begin{array}{c} m+p+k+1, -l-q-w; \\ \times m+p+k+2; \\ \times t(K_d+1)/(t(K_d+1)+S(K_i+1)) \end{array}\right)}{\left(t(K_d+1)+S(K_i+1)\right)^{2m+2n+l+k+p+q+3}(m+p+k+1)}\right].$$

$$G_{25} = G_{24}(m+p+k+1)(m+n+k+1)$$

$$\times (K_d+1)^{2k+2m+n+p+2}(K_i+1)^{l+q+1}. \tag{6.57}$$

The PDF of the SC output SIR for various values of the correlation coefficient and Rice factor is graphically presented in Figure 6.29.

OP versus normalized parameter S/γ_{th} for various values of parameters ρ and K_d is shown in Figure 6.30. It can be observed from this figure that the OP deteriorates with a decrease in parameter K_d. Also, the presented results show branch correlation influence on the OP. When correlation coefficients decrease, the OP increases. In Figure 6.30, it is pointed out that the influence of fading severity variety on outage performance depends on the correlation coefficient and it is even less if correlation coefficient is greater.

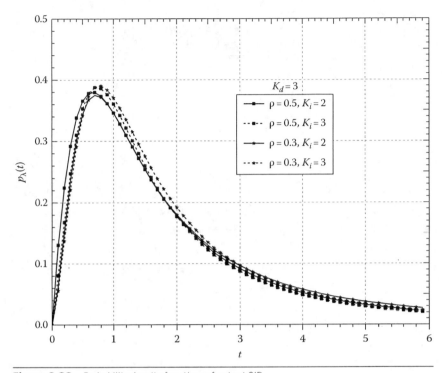

Figure 6.29 Probability density functions of output SIR.

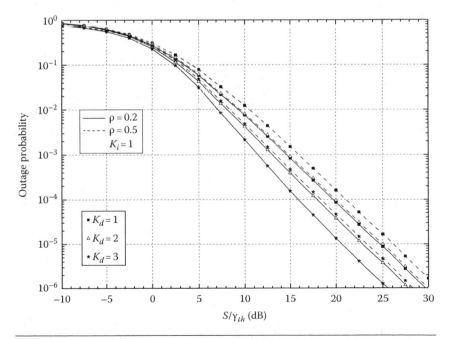

Figure 6.30 OP versus S/γ_{th} at the output of dual SIR-based SC with correlated branches.

Table 6.4   Number of Terms of
(6.56) Required for Three Significant
Figure Accuracy ($S/\gamma_{th} = 0$ dB)

ρ	$K_d=1$ $K_i=1$	$K_d=2$ $K_i=1$	$K_d=3$ $K_i=1$
0.2	5	6	8
0.5	9	12	17

Convergence of the nested infinite sum in (6.56) is shown in Table 6.4. Expressions (6.56) converge for any value of the parameters ρ, K_d, and K_i, but as shown in Table 6.4, the number of the terms needed to be summed to achieve the desired accuracy depends strongly on the correlation coefficient ρ and the Rice factor values. Actually, the number of the terms increases with an increase in both correlation coefficient and Rice factor values.

ASEP is numerically obtained and shown in Figure 6.31 for some values of ρ and K_d. It can be concluded from this figure that the system shows better error performances for a larger distance

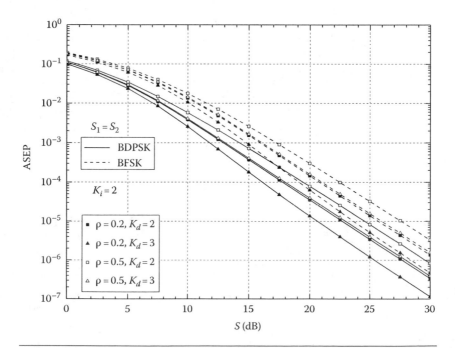

Figure 6.31   ASEP versus average input SIR for BDPSK and noncoherent BFSK system.

between diversity branches, that is, for a lower value of correlation coefficient. This is more notable for both greater values of the average SIR and higher values of the Rice factor of the desired signal. Results from Figure 6.31 also present the influence of fading severity on ASEP. ASEP decreases with the decrease in fading severity. Comparison of curves from Figure 6.31 shows better error performance of the BDPSK modulation technique than the noncoherent BFSK system, especially when the desired signal dominates in relation to interference.

6.3 Diversity Reception over Generalized K Fading Channels in the Presence of CCI

6.3.1 SSC Diversity Reception with Uncorrelated Branches

Applying a similar approach as in Sections 6.1.1 and 6.1.2, starting from Equations 6.1 and 6.2 and considering SIR values at antennas, λ_i, statistically independent, the PDF of the output SIR, λ, can be presented, capitalizing on

$$f_{\lambda_1}(\lambda) = \lambda^{(m_{d1}+k_{d1})/2-1}\left(\frac{m_{d1}}{m_{c1}S_1}\right)^{(m_{d1}+k_{d1})/2}\frac{1}{\Gamma(m_{d1})\Gamma(k_{d1})\Gamma(m_{c1})\Gamma(k_{c1})}$$

$$\times G_{2,2}^{2,2}\left[m_{d1}\lambda/m_{c1}S_1\left|\begin{array}{l}1-(m_{d1}+k_{d1}+2k_{c1})/2,\\ 1-(m_{d1}+k_{d1}+2m_{c1})/2\\ (k_{d1}-m_{d1})/2,-(k_{d1}-m_{d1})/2\end{array}\right.\right];$$

$$f_{\lambda_2}(\lambda) = \lambda^{(m_{d2}+k_{d2})/2-1}\left(\frac{m_{d2}}{m_{c2}S_2}\right)^{(m_{d2}+k_{d2})/2}\frac{1}{\Gamma(m_{d2})\Gamma(k_{d2})\Gamma(m_{c2})\Gamma(k_{c2})}$$

$$\times G_{2,2}^{2,2}\left[m_{d2}\lambda/m_{c2}S_2\left|\begin{array}{l}1-(m_{d2}+k_{d2}+2k_{c2})/2,\\ 1-(m_{d2}+k_{d2}+2m_{c2})/2\\ (k_{d2}-m_{d2})/2,-(k_{d2}-m_{d2})/2\end{array}\right.\right];$$

$$F_{\lambda_1}(z_T)$$

$$= \int_0^{z_T} f_{\lambda_1}(\lambda)\, d\lambda = \frac{z_T^{(m_{d1}+k_{d1})/2}}{\Gamma(m_{d1})\Gamma(k_{d1})\Gamma(m_{c1})\Gamma(k_{c1})} (m_{d1}/m_{c1}S_1)^{(m_{d1}+k_{d1})/2}$$

$$\times G_{3,3}^{2,3}\left[m_{d1}z_T/m_{c1}S_2 \,\middle|\, \begin{array}{l} 1-(m_{d1}+k_{d1}+2k_{c1})/2, \\[4pt] 1-(m_{d1}+k_{d1}+2m_{c1})/2, \\[4pt] 1-(m_{d1}+k_{d1})/2 \\[8pt] (k_{d1}-m_{d1})/2, -(k_{d1}-m_{d1})/2, \\[4pt] -(m_{d1}+k_{d1})/2 \end{array} \right];$$

$$F_{\lambda_2}(z_T)$$

$$= \int_0^{z_T} f_{\lambda_2}(\lambda)\, d\lambda = \frac{z_T^{(m_{d2}+k_{d2})/2}}{\Gamma(m_{d2})\Gamma(k_{d2})\Gamma(m_{c2})\Gamma(k_{c2})} (m_{d2}/m_{c2}S_2)^{(m_{d2}+k_{d2})/2}$$

$$\times G_{3,3}^{2,3}\left[m_{d2}z_T/m_{c2}S_2 \,\middle|\, \begin{array}{l} 1-(m_{d2}+k_{d2}+2k_{c2})/2, \\[4pt] 1-(m_{d2}+k_{d2}+2m_{c2})/2, \\[4pt] 1-(m_{d2}+k_{d2})/2 \\[8pt] (k_{d2}-m_{d2})/2, -(k_{d2}-m_{d2})/2, \\[4pt] -(m_{d2}+k_{d2})/2 \end{array} \right], \qquad (6.58)$$

where

$G_{p,q}^{m,n}\left[\,.\,|\,.\,\right]$ is the Meijer's G-function [8, Eq. (9.301)]

S_i stands for the average SIRs at the two input branches, $i=1,2$, defined as $S_i = \Omega_{di}/\Omega_{ci}$, $(\Omega_{di}=E(R_i^2)/k_{di}$, $\Omega_{ci}=E(r_i^2)/k_{ci})$

Similarly, the CDF at the output of the SSC combiner can be determined capitalizing on

$$F_{\lambda_1}(\lambda) = \frac{\lambda^{(m_{d1}+k_{d1})/2}}{\Gamma(m_{d1})\Gamma(k_{d1})\Gamma(m_{c1})\Gamma(k_{c1})}(m_{d1}/m_{c1}S_1)^{(m_{d1}+k_{d1})/2}$$

$$\times G_{3,3}^{2,3}\left[m_{d1}\lambda/m_{c1}S_1 \left| \begin{array}{c} 1-(m_{d1}+k_{d1}+2k_{c1})/2, \\[1em] 1-(m_{d1}+k_{d1}+2m_{c1})/2, \\[1em] 1-(m_{d1}+k_{d1})/2 \\[1em] (k_{d1}-m_{d1})/2,-(k_{d1}-m_{d1})/2, \\[1em] -(m_{d1}+k_{d1})/2 \end{array}\right.\right];$$

$$F_{\lambda_2}(\lambda) = \frac{\lambda^{(m_{d2}+k_{d2})/2}}{\Gamma(m_{d2})\Gamma(k_{d2})\Gamma(m_{c2})\Gamma(k_{c2})}(m_{d2}/m_{c2}S_2)^{(m_{d2}+k_{d2})/2}$$

$$\times G_{3,3}^{2,3}\left[m_{d2}\lambda/m_{c2}S_2 \left| \begin{array}{c} 1-(m_{d2}+k_{d2}+2k_{c2})/2, \\[1em] 1-(m_{d2}+k_{d2}+2m_{c2})/2, \\[1em] 1-(m_{d2}+k_{d2})/2 \\[1em] (k_{d2}-m_{d2})/2,-(k_{d2}-m_{d2})/2, \\[1em] -(m_{d2}+k_{d2})/2 \end{array}\right.\right]. \qquad (6.59)$$

Based on the presented statistics, system performance measures could be efficiently evaluated.

6.3.2 SSC Diversity Reception with Correlated Branches

Assuming that because of the insufficient antenna spacing, both desired and interfering signal envelopes experience correlative Generalized K fading with joint distributions [24,25]

$$f_{R_1 R_2}\left(R_1, R_2\right) = \frac{16}{\Gamma\left(m_d\right)\Gamma\left(k_d\right)} \sum_{a,b=0}^{\infty} \frac{m_d^{\varepsilon_d}\,\rho_{nd}^{a}\,\rho_{gd}^{b}}{\Gamma\left(m_d + a\right)\Gamma\left(k_d + b\right)}$$

$$\times \frac{\prod_{l=1}^{2}\left(R_l / \sqrt{\Omega_{dl}}\right)^{\varepsilon_d} K_{\psi}\left(2R_l \sqrt{m_d /\left(1-\rho_{nd}\right)\left(1-\rho_{gd}\right)\Omega_{dl}}\right)}{a!\,b!\left(1-\rho_{nd}\right)^{k_d+a+b}\left(1-\rho_{gd}\right)^{m_d+a+b} R_1 R_2}$$

$$(6.60)$$

and

$$f_{r_1 r_2}\left(r_1, r_2\right) = \frac{16}{\Gamma\left(m_c\right)\Gamma\left(k_c\right)} \sum_{c,d=0}^{\infty} \frac{m_c^{\varepsilon_c}\,\rho_{nc}^{c}\,\rho_{gc}^{d}}{\Gamma\left(m_c + c\right)\Gamma\left(k_c + d\right)}$$

$$\times \frac{\prod_{l=1}^{2}\left(r_l / \sqrt{\Omega_{cl}}\right)^{\varepsilon_c} K_{\psi}\left(2r_l \sqrt{m_c /\left(1-\rho_{nc}\right)\left(1-\rho_{gc}\right)\Omega_{cl}}\right)}{c!\,d!\left(1-\rho_{nc}\right)^{k_c+c+d}\left(1-\rho_{gc}\right)^{m_c+c+d} r_1 r_2},$$

$$(6.61)$$

where

ρ_{nd} is the power correlation coefficient between instantaneous powers of Nakagami-m fading processes and corresponds to desired signals

ρ_{gd} is the correlation coefficient between the average gamma fading powers of desired signals

ρ_{nc} and ρ_{gc} represent correlation levels related to the fading/shadowing interfering signal

Ω_{dl} and Ω_{cl} denote the average powers of the desired and interfering signals, affected by the composite generalized K fading/shadowing model

$z_1 = R_1^2 / r_1^2$ and $z_2 = R_2^2 / r_2^2$ represent the instantaneous SIR on the diversity branches, respectively

After substituting (6.60) and (6.61) in (6.8), we obtain the joint PDF of the SIR in the form of

$$f_{z_1 z_2}(z_1, z_2) = \sum_{a,b,c,d=0}^{+\infty} G_{26} \times z_1^{\varepsilon_d/2-1} z_2^{\varepsilon_d/2-1}$$

$$\times G_{2,2}^{2,2} \left[m_d z_1 \sigma/m_c S \left| \begin{array}{l} 1-(\varepsilon_d+\varepsilon_c+\psi_c)/2, \\ 1-(\varepsilon_d+\varepsilon_c-\psi_c)/2 \\ \psi_d/2, -\psi_d/2 \end{array} \right. \right]$$

$$\times G_{2,2}^{2,2} \left[m_d z_2 \sigma/m_c S \left| \begin{array}{l} 1-(\varepsilon_d+\varepsilon_c+\psi_c)/2, \\ 1-(\varepsilon_d+\varepsilon_c-\psi_c)/2 \\ \psi_d/2, -\psi_d/2 \end{array} \right. \right]$$

$$G_{26} = \frac{m_d^{\varepsilon_d} \rho_{nd}^a \rho_{gd}^b}{\Gamma(m_d)\Gamma(k_d)\Gamma(m_c)\Gamma(k_c) m_c^{\varepsilon_d} \Gamma(m_d+a)}$$

$$\times \Gamma(k_d+b)\Gamma(m_c+c)\Gamma(k_c+d) S^{\varepsilon_d} a! b! c! d!$$

$$\times \frac{\rho_{nc}^c \rho_{gc}^d}{(1-\rho_{nd})^{k_d+a+b}(1-\rho_{gd})^{m_d+a+b}};$$

$$\times (1-\rho_{nc})^{k_c+c+d-\varepsilon_d-\varepsilon_c}(1-\rho_{gc})^{m_c+c+d-\varepsilon_d-\varepsilon_c}$$

$$\sigma = (1-\rho_{nc})(1-\rho_{gc})/(1-\rho_{nd})(1-\rho_{gd}); \quad \varepsilon_d = k_d+m_d+a+b;$$

$$\psi_d = k_d+b-m_d-a; \quad \psi_c = k_c+d-m_c-c, \qquad (6.62)$$

where

$S_k = \Omega_{dk}/\Omega_{ck}$ is the average SIR at the kth input branch of balanced SC ($S_1 = S_2 = S$)

z represents the instantaneous SIR at the SSC output

z_T is the predetermined switching threshold for both input branches

The output SSC PDF can be obtained according to (6.11) and (6.12), and for this case $v_{SSC}(z)$ can be expressed as in closed form after some basic transformations:

$$v_{SSC}(z) = \int_0^{z_T} f_{z_1 z_2}(z, z_2) dz_2 = \sum_{a,b,c,d=0}^{+\infty} G_{26} \times z^{\varepsilon_d/2-1} z_T^{\varepsilon_d/2}$$

$$\times G_{2,2}^{2,2} \left[m_d z \sigma / m_c S \left| \begin{array}{c} 1-(\varepsilon_d + \varepsilon_c + \psi_c)/2, \\ 1-(\varepsilon_d + \varepsilon_c - \psi_c)/2 \\ \psi_d/2, -\psi_d/2 \end{array} \right. \right]$$

$$\times G_{3,3}^{2,3} \left[m_d z_T \sigma / m_c S \left| \begin{array}{c} 1-(\varepsilon_d + \varepsilon_c + \psi_c)/2, \\ 1-(\varepsilon_d + \varepsilon_c - \psi_c)/2, \\ (1-\varepsilon_d)/2 \\ \psi_d/2, \ -\psi_d/2, -\varepsilon_d/2 \end{array} \right. \right]. \quad (6.63)$$

Assuming that due to propagation conditions $m_{d1} = m_{d2} = m_d$, $m_{c1} = m_{c2} = m_c$, $k_{d1} = k_{d2} = k_d$, $k_{c1} = k_{c2} = k_c$, and $S_1 = S_2 = S$, while in the similar manner

$$f_{z_1}(z) = z^{(m_d + k_d)/2-1} \left(\frac{m_d}{m_c S_1} \right)^{(m_d + k_d)/2} \frac{1}{\Gamma(m_d)\Gamma(k_d)\Gamma(m_c)\Gamma(k_c)}$$

$$\times G_{2,2}^{2,2} \left[m_d z / m_c S_1 \left| \begin{array}{c} 1-(m_d + k_d + 2k_c)/2, 1-(m_d + k_d + 2m_c)/2 \\ (k_d - m_d)/2, -(k_d - m_d)/2 \end{array} \right. \right].$$

$$(6.64)$$

CDF of instantaneous output SIR can be expressed according to (6.4) through $F_z(z)$, $F_z(z_T)$, and $F_{z_1 z_2}(z, z_T)$ derived in the forms of

$$F_{z_1z_2}\left(z,z_T\right)=\int_0^z\int_0^{z_T} f_{z_1z_2}\left(z_1,z_2\right)dz_1dz_2=\sum_{a,b,c,d=0}^{+\infty} G_{26}\times z^{\varepsilon_d/2}z_T^{\varepsilon_d/2}$$

$$\times G_{3,3}^{2,3}\left[m_d z\sigma/m_c S\,\middle|\,\begin{matrix}1-\left(\varepsilon_d+\varepsilon_c+\psi_c\right)/2,\\ 1-\left(\varepsilon_d+\varepsilon_c-\psi_c\right)/2,\\ 1-\varepsilon_d/2\\ \psi_d/2,\ -\psi_d/2,-\varepsilon_d/2\end{matrix}\right]$$

$$\times G_{3,3}^{2,3}\left[m_d z_T\sigma/m_c S\,\middle|\,\begin{matrix}1-\left(\varepsilon_d+\varepsilon_c+\psi_c\right)/2,\\ 1-\left(\varepsilon_d+\varepsilon_c-\psi_c\right)/2,\\ 1-\varepsilon_d/2\\ \psi_d/2,\ -\psi_d/2,-\varepsilon_d/2\end{matrix}\right];$$

$$F_z\left(z\right)=\frac{z^{(m_d+k_d)/2}}{\Gamma\left(m_d\right)\Gamma\left(k_d\right)\Gamma\left(m_c\right)\Gamma\left(k_c\right)}\left(m_d/m_c S\right)^{(m_d+k_d)/2}$$

$$\times G_{3,3}^{2,3}\left[m_d z/m_c S\,\middle|\,\begin{matrix}1-\left(m_d+k_d+2k_c\right)/2,\\ 1-\left(m_d+k_d+2m_c\right)/2,\\ 1-\left(m_d+k_d\right)/2\\ \left(k_d-m_d\right)/2,-\left(k_d-m_d\right)/2,\\ -\left(m_d+k_d\right)/2\end{matrix}\right];$$

$$F_z\left(z\right)=\frac{z_T^{(m_d+k_d)/2}}{\Gamma\left(m_d\right)\Gamma\left(k_d\right)\Gamma\left(m_c\right)\Gamma\left(k_c\right)}\left(m_d/m_c S\right)^{(m_d+k_d)/2}$$

$$\times G_{3,3}^{2,3}\left[m_d z_T/m_c S\,\middle|\,\begin{matrix}1-\left(m_d+k_d+2k_c\right)/2,\\ 1-\left(m_d+k_d+2m_c\right)/2,\\ 1-\left(m_d+k_d\right)/2\\ \left(k_d-m_d\right)/2,-\left(k_d-m_d\right)/2,\\ -\left(m_d+k_d\right)/2\end{matrix}\right]. \tag{6.65}$$

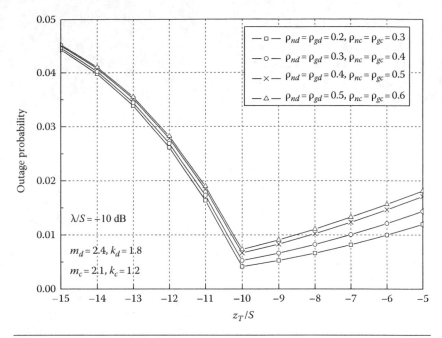

Figure 6.32   OP versus normalized switching threshold for various correlation coefficient values.

Figure 6.32 shows the OP versus the normalized switching threshold z_T/S for various correlation coefficient values. It is obvious that there is an optimal threshold that minimizes the OP. For this threshold value, SSC combining can be observed as SC.

The OP versus normalized average SIR for different values of shaping parameters is shown in Figure 6.33. It is interesting to note here that for pre-Rayleigh fading conditions ($m_d = 0.5$, $m_c = 0.4$), the system performance is not acceptable, even for light shadowing condition ($k_d = 5.3$, $k_c = 1.1$). The performance improves significantly when small-scale fading effects decrease ($m_d = 3.8$, $m_c = 0.4$). In particular, the OP is predominantly affected by the fading parameters of the desired user, rather than by the fading parameters of the interferers.

6.3.3 SC Diversity Reception with Uncorrelated Branches

The CDF and PDF of the SC output SIR can be determined following a similar procedure as in Sections 6.1.3 and 6.2.3, and they can be presented in the following forms:

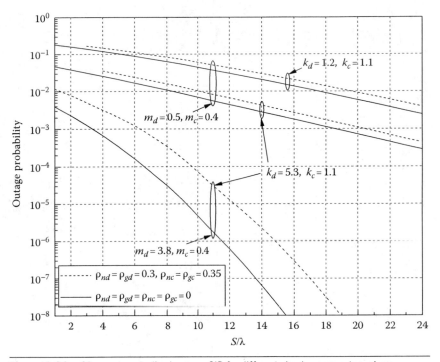

Figure 6.33   OP versus normalized average SIR for different shaping parameter values.

$$F_{\lambda}(t) = F_{\lambda_1}(t)F_{\lambda_2}(t)\cdots F_{\lambda_N}(t) = \prod_{i=1}^{N} F_{\lambda_i}(t)$$

$$= \prod_{i=1}^{N}\left\{\frac{t^{(m_{di}+k_{di})/2}}{\Gamma(m_{di})\Gamma(k_{di})\Gamma(m_{ci})\Gamma(k_{ci})}(m_{di}/m_{ci}S_i)^{(m_{di}+k_{di})/2}\right.$$

$$\left.\times G_{3,3}^{2,3}\left[m_{di}t/m_{ci}S_i \left|\begin{array}{c}1-(m_{di}+k_{di}+2k_{ci})/2, \\ 1-(m_{di}+k_{di}+2m_{ci})/2, \\ 1-(m_{di}+k_{di})/2 \\ (k_{di}-m_{di})/2, -(k_{di}-m_{di})/2, \\ -(m_{di}+k_{di})/2\end{array}\right.\right]\right\}$$

$$\tag{6.66}$$

and

$$f_\lambda(\lambda) = \frac{d}{d\lambda} F(\lambda) = \sum_{i=1}^{n} f_{\lambda_i}(\lambda) \prod_{\substack{j=1 \\ j\neq i}}^{n} F_{\lambda_i}(\lambda)$$

$$= \left(\frac{m_{di}}{m_{ci}S_i}\right)^{(m_{di}+k_{di})/2} \frac{\lambda^{(m_{di}+k_{di})/2-1}}{\Gamma(m_{di})\Gamma(k_{di})\Gamma(m_{ci})\Gamma(k_{ci})}$$

$$\times G_{2,2}^{2,2}\left[m_{di}\lambda/m_{ci}S_i \left| \begin{array}{c} 1-(m_{di}+k_{di}+2k_{ci})/2, 1-(m_{di}+k_{di}+2m_{ci})/2 \\ (k_{di}-m_{di})/2, -(k_{di}-m_{di})/2 \end{array} \right. \right]$$

$$\times \prod_{\substack{j=1 \\ j\neq i}}^{n} \left\{ \begin{array}{l} \dfrac{\lambda^{(m_{dj}+k_{dj})/2}}{\Gamma(m_{dj})\Gamma(k_{dj})\Gamma(m_{cj})\Gamma(k_{cj})}(m_{dj}/m_{cj}S_j)^{(m_{dj}+k_{dj})/2} \\ \\ \times G_{3,3}^{2,3}\left[m_{dj}\lambda/m_{cj}S_j \left| \begin{array}{l} 1-(m_{dj}+k_{dj}+2k_{cj})/2, \\ 1-(m_{dj}+k_{dj}+2m_{cj})/2, \\ 1-(m_{dj}+k_{dj})/2 \\ (k_{dj}-m_{dj})/2, -(k_{dj}-m_{dj})/2, \\ -(m_{dj}+k_{dj})/2 \end{array} \right. \right] \end{array} \right\}. $$

$$(6.67)$$

Instantaneous and average values of the SIR at the kth input branch of SC combiner are denoted with λ_k and S_k in previous relations.

6.3.4 SC Diversity Reception with Correlated Branches

Let us now consider a dual-branch SIR-based SC receiver with correlated branches in this section. Let $S_k = \Omega_{dk}/\Omega_{ck}$ denote the average SIR at the kth input branch of the SC. The CDF of the output SIR can be derived from (6.62) by using (6.33) and equating the arguments $t_1 = t_2 = t$ as

$$F_\lambda(t) = \frac{1}{\Gamma(m_d)\Gamma(k_d)\Gamma(m_c)\Gamma(k_c)}$$

$$\times \sum_{i,j,i_c,j_c=0}^{+\infty} \frac{m_d^{\xi_d} \rho_{Nd}^i \rho_{Gd}^j m_c^{-\xi_d} \rho_{Nc}^{i_c} \rho_{Gc}^{j_c}}{\Gamma(m_d+i)\Gamma(k_d+j)\Gamma(m_c+i_c)\Gamma(k_c+j_c)}$$

$$\times \frac{t^{\xi_d}}{i!\,j!\,(1-\rho_{Nd})^{k_d+i+j}\,(1-\rho_{Gd})^{m_d+i+j}}$$

$$\times \frac{(S_1 S_2)^{-\xi_d/2}}{i_c!\,j_c!\,(1-\rho_{Nc})^{k_c+i_c+j_c-\xi_d-\xi_c}\,(1-\rho_{Gc})^{m_c+i_c+j_c-\xi_d-\xi_c}}$$

$$\times G_{3,3}^{2,3}\left(\frac{m_d(1-\rho_{Nc})(1-\rho_{Gc})t}{m_c(1-\rho_{Nd})(1-\rho_{Gd})S_1}\left|\begin{array}{c} 1-\dfrac{\xi_d+\xi_c+\psi_c}{2},\\[2mm] 1-\dfrac{\xi_d+\xi_c-\psi_c}{2},1-\dfrac{\xi_d}{2}\\[2mm] \psi_d/2,-\psi_d/2,-\xi_d/2 \end{array}\right.\right)$$

$$\times G_{3,3}^{2,3}\left(\frac{m_d(1-\rho_{Nc})(1-\rho_{Gc})t}{m_c(1-\rho_{Nd})(1-\rho_{Gd})S_2}\left|\begin{array}{c} 1-\dfrac{\xi_d+\xi_c+\psi_c}{2},\\[2mm] 1-\dfrac{\xi_d+\xi_c-\psi_c}{2},1-\dfrac{\xi_d}{2}\\[2mm] \psi_d/2,-\psi_d/2,-\xi_d/2 \end{array}\right.\right).$$

$$(6.68)$$

Using (6.68), one can easily obtain a PDF of the output SIR as

$$f_\lambda(t) = \frac{d}{dt} F_\lambda(t) = \frac{1}{\Gamma(m_d)\Gamma(k_d)\Gamma(m_c)\Gamma(k_c)}$$

$$\times \sum_{i,j,i_c,j_c=0}^{+\infty} \frac{m_d^{\xi_d}\,\rho_{Nd}^i\,\rho_{Gd}^j\,m_c^{-\xi_d}\,\rho_{Nc}^{i_c}\,\rho_{Gc}^{j_c}}{\Gamma(m_d+i)\Gamma(k_d+j)\Gamma(m_c+i_c)\Gamma(k_c+j_c)}$$

$$\times \frac{t^{\xi_d-1}}{i!\,j!\,(1-\rho_{Nd})^{k_d+i+j}\,(1-\rho_{Gd})^{m_d+i+j}}$$

$$\times \frac{(S_1 S_2)^{-\xi_d/2}}{i_c!\,j_c!\,(1-\rho_{Nc})^{k_c+i_c+j_c-\xi_d-\xi_c}\,(1-\rho_{Gc})^{m_c+i_c+j_c-\xi_d-\xi_c}}$$

$$\times \left(\xi_d G_{3,3}^{2,3} \left(\frac{m_d \left(1-\rho_{Nc}\right)\left(1-\rho_{Gc}\right)t}{m_c \left(1-\rho_{Nd}\right)\left(1-\rho_{Gd}\right)S_1} \middle| \begin{matrix} 1-\dfrac{\xi_d+\xi_c+\psi_c}{2}, \\[2mm] 1-\dfrac{\xi_d+\xi_c-\psi_c}{2}, \\[2mm] 1-\dfrac{\xi_d}{2} \\[2mm] \psi_d/2,-\psi_d/2,-\xi_d/2 \end{matrix} \right) \right.$$

$$\times G_{3,3}^{2,3} \left(\frac{m_d \left(1-\rho_{Nc}\right)\left(1-\rho_{Gc}\right)t}{m_c \left(1-\rho_{Nd}\right)\left(1-\rho_{Gd}\right)S_2} \middle| \begin{matrix} 1-\dfrac{\xi_d+\xi_c+\psi_c}{2}, \\[2mm] 1-\dfrac{\xi_d+\xi_c-\psi_c}{2}, \\[2mm] 1-\dfrac{\xi_d}{2} \\[2mm] \psi_d/2,-\psi_d/2,-\xi_d/2 \end{matrix} \right)$$

$$+ G_{4,4}^{2,4} \left(\frac{m_d \left(1-\rho_{Nc}\right)\left(1-\rho_{Gc}\right)t}{m_c \left(1-\rho_{Nd}\right)\left(1-\rho_{Gd}\right)S_1} \middle| \begin{matrix} 0,1-\dfrac{\xi_d+\xi_c+\psi_c}{2}, \\[2mm] 1-\dfrac{\xi_d+\xi_c-\psi_c}{2}, \\[2mm] 1-\dfrac{\xi_d}{2} \\[2mm] \psi_d/2,-\psi_d/2,1,-\xi_d/2 \end{matrix} \right)$$

$$\times G_{3,3}^{2,3} \left(\frac{m_d \left(1-\rho_{Nc}\right)\left(1-\rho_{Gc}\right)t}{m_c \left(1-\rho_{Nd}\right)\left(1-\rho_{Gd}\right)S_2} \middle| \begin{matrix} 1-\dfrac{\xi_d+\xi_c+\psi_c}{2}, \\[2mm] 1-\dfrac{\xi_d+\xi_c-\psi_c}{2}, \\[2mm] 1-\dfrac{\xi_d}{2} \\[2mm] \psi_d/2,-\psi_d/2,-\xi_d/2 \end{matrix} \right)$$

$$\times G_{3,3}^{2,3} \left(\frac{m_d \left(1-\rho_{Nc}\right)\left(1-\rho_{Gc}\right)t}{m_c \left(1-\rho_{Nd}\right)\left(1-\rho_{Gd}\right)S_1} \left| \begin{matrix} 1 - \dfrac{\xi_d + \xi_c + \psi_c}{2}, \\[2mm] 1 - \dfrac{\xi_d + \xi_c - \psi_c}{2}, \\[2mm] 1 - \dfrac{\xi_d}{2} \\[2mm] \psi_d/2, -\psi_d/2, -\xi_d/2 \end{matrix} \right. \right)$$

$$\times G_{4,4}^{2,4} \left(\frac{m_d \left(1-\rho_{Nc}\right)\left(1-\rho_{Gc}\right)t}{m_c \left(1-\rho_{Nd}\right)\left(1-\rho_{Gd}\right)S_2} \left| \begin{matrix} 0, 1 - \dfrac{\xi_d + \xi_c + \psi_c}{2}, \\[2mm] 1 - \dfrac{\xi_d + \xi_c - \psi_c}{2}, \\[2mm] 1 - \dfrac{\xi_d}{2} \\[2mm] \psi_d/2, -\psi_d/2, 1, -\xi_d/2 \end{matrix} \right. \right) .$$

$$(6.69)$$

In Figure 6.34, the dependence of the OP on normalized instantaneous SIR λ/S is presented for two values of the Nakagami-m shaping parameter for the desired signal m_d. The increase in m_d value causes OP to decrease, and this effect is emphasized more for small values of λ/S. The influence of the interference signal Nakagami-m fading parameter m_c on the OP is negligible when other parameters are chosen as in Figure 6.34.

The shadowing parameter of interfering signal k_c dramatically affects the OP, so the increase in this value significantly impairs system performance. However, the shadowing parameter of the desired signal k_d, among all the other parameters of composite fading shows the largest impact on the system performance, as illustrated in Figure 6.35 (lower OP corresponds to larger k_d values). In Figure 6.36, the influence of correlation coefficients between desired signals ρ_{Nd} and ρ_{Gd} on the OP is analyzed. For small λ/S values, one can notice that lower values of ρ_{Gd} improve system characteristics, while this effect is much less significant for larger λ/S values.

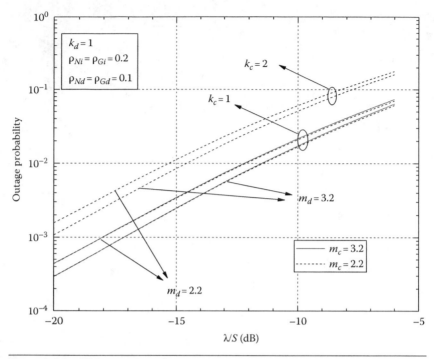

Figure 6.34   OP for two values of parameter m_d.

The influence of the ρ_{Nd} coefficient stays equally emphasized for a wide range of λ/S values. In Figure 6.37, the influence of correlation coefficients between interfering signals ρ_{Nc} and ρ_{Gc} on the OP is presented. The influence of the ρ_{Nc} coefficient is almost negligible, while larger ρ_{Gc} values lead to an increase in OP.

6.4 Diversity Reception over Rayleigh Fading Channels in Experiencing an Arbitrary Number of Multiple CCI

Assume that there are M_i independent equal power Rayleigh distributed interferers over the ith branch of the SC diversity system with an arbitrary number of branches. The identical local mean power interferers assumption has been adopted by many other authors [26–29]. Alouini et al. have driven the conclusion in [28] that equal average power interferers assumption is suitable for the two limiting cases that can bound the performances of any interference-limited systems. These two limited cases correspond to the scenario when the interferers are on the cell edges closest to

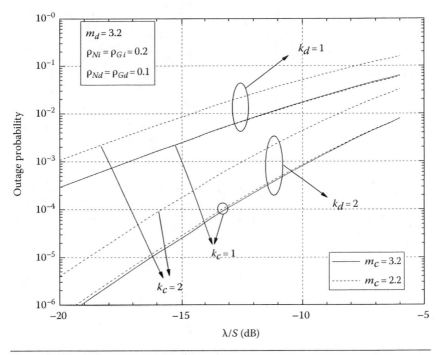

Figure 6.35 OP for two values of parameter k_d.

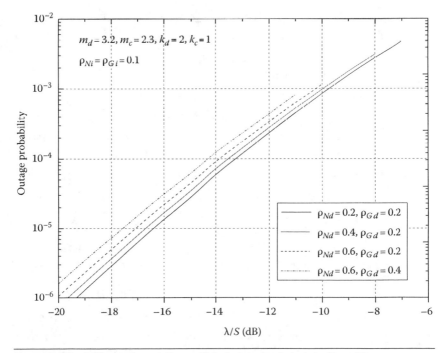

Figure 6.36 Influence of correlation coefficients between desired signals on OP.

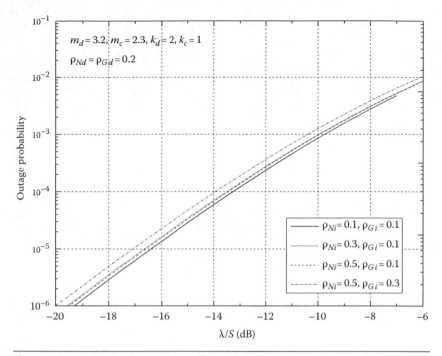

Figure 6.37   Influence of correlation coefficients between interfering signals on OP.

the desired user cell (worst interference case scenario) or where they are at the furthest edges (best interference case scenario). Finally, in practice, there are several wireless systems that could be adequately modeled using the assumption of equal average power interferers as explained in [26–28], such as a single multiantenna interferer or an interfering cluster of colocated terminals. For example [27], if we are considering one MIMO interferer, with closely spaced antennas, interfering signals coming from different antennas are subject to statistically identical fading processes. The independent instantaneous interfering signals are added together to produce the resultant instantaneous interfering signal at the ith branch of the diversity system and can be written as

$$I_i = I_{1i} + I_{2i} + I_{3i} + \cdots + I_{M_i}. \tag{6.70}$$

Since each interfering signal amplitude is modeled with the Rayleigh distribution, the SNR (signal-to-noise ratio) PDF of the sum of M_i Rayleigh distributed random variables (PDF of the SNR of the total interfering signal) is then approximated by [30]

$$f_{y_i}(y_i) = \frac{y_i^{M_i-1}\exp(-y_i/\Omega_{yi})}{\Omega_{yi}^{M_i}(M_i-1)!}; \quad \Omega_{yi} = \frac{2\Omega_{0yi}}{M_i}\left(\frac{\Gamma(M_i+1/2)}{\Gamma(1/2)}\right)^{1/M_i},$$

$$(6.71)$$

where Ω_{yi} is the total interference power at the ith branch of the diversity system, given in the function of the average power of each interferer Ω_{0yi}.

The desired signal envelopes on the ith diversity branch also follow the Rayleigh fading distribution, whose PDF is given by

$$f_{x_i}(x_i) = \frac{1}{\Omega_{xi}}\exp\left(\frac{x_i}{\Omega_{xi}}\right). \qquad (6.72)$$

Let $\lambda_i = x_i/y_i$ be the SIR at the ith $(i = 1, 2, \ldots, N)$ diversity branch of the receiver. Then PDF of λ_i can be written as

$$f_{\lambda_i}(\lambda_i) = \int_0^\infty y_i f_{x_i}(y_i\lambda_i) f_{y_i}(y_i)dy_i;$$

$$f_{\lambda_i}(\lambda_i) = \frac{\left(\dfrac{S_iM_i}{2\left(\Gamma(M_i+1/2)/\Gamma(1/2)\right)^{M_i}}\right)^{M_i}\Gamma(M_i+1)}{(M_i-1)!\left(\lambda_i+\dfrac{S_iM_i}{2\left(\Gamma(M_i+1/2)/\Gamma(1/2)\right)^{M_i}}\right)^{M_i+1}}, \quad (6.73)$$

with $S_i = \Omega_{xi}/\Omega_{0yi}$ being the average SIRs at the ith input branch of the diversity system. Similarly, CDF can be after some mathematical manipulation written in the form of

$$F_{\lambda_i}(\lambda_i) = \int_0^{\lambda_i} f_{t_i}(t_i)dt_i;$$

$$F_{\lambda_i}(\lambda_i) = \left(\lambda_i+\frac{S_iM_i}{2\left(\Gamma(M_i+1/2)/\Gamma(1/2)\right)^{M_i}}\right)\frac{\Gamma(M_i+1)}{(M_i-1)!}$$

$$\times {}_2F_1\left(1,2;1-M_i;\lambda_i+\frac{S_iM_i}{2\left(\Gamma(M_i+1/2)/\Gamma(1/2)\right)^{M_i}}\right) \qquad (6.74)$$

6.4.1 SSC Diversity Reception with Uncorrelated Branches

Capitalizing on Equations 6.73 and 6.74 with respect to (6.1) and (6.2), the PDF and CDF of the output SSC SIR, λ, can be presented as

$$
f_{\lambda_1}(\lambda) = \frac{\left(\dfrac{S_1 M_1}{2\left(\Gamma(M_1 + 1/2)/\Gamma(1/2)\right)^{M_1}}\right)^{M_1} \Gamma(M_1 + 1)}{(M_1 - 1)!\left(\lambda + \dfrac{S_1 M_1}{2\left(\Gamma(M_i + 1/2)/\Gamma(1/2)\right)^{M_1}}\right)^{M_1+1}};
$$

$$
f_{\lambda_2}(\lambda) = \frac{\left(\dfrac{S_2 M_2}{2\left(\Gamma(M_2 + 1/2)/\Gamma(1/2)\right)^{M_1}}\right)^{M_2} \Gamma(M_2 + 1)}{(M_2 - 1)!\left(\lambda + \dfrac{S_1 M_2}{2\left(\Gamma(M_i + 1/2)/\Gamma(1/2)\right)^{M_2}}\right)^{M_2+1}}
$$

$$
F_{\lambda_1}(z_T) = \left(z_T + \dfrac{S_1 M_1}{2\left(\Gamma(M_1 + 1/2)/\Gamma(1/2)\right)^{M_1}}\right) \dfrac{\Gamma(M_1 + 1)}{(M_1 - 1)!}
$$

$$
\times {}_2F_1\left(1, 2; 1 - M_1; z_T + \dfrac{S_2 M_1}{2\left(\Gamma(M_1 + 1/2)/\Gamma(1/2)\right)^{M_1}}\right);
$$

$$
F_{\lambda_2}(z_T) = \left(z_T + \dfrac{S_2 M_2}{2\left(\Gamma(M_2 + 1/2)/\Gamma(1/2)\right)^{M_2}}\right) \dfrac{\Gamma(M_2 + 1)}{(M_2 - 1)!}
$$

$$
\times {}_2F_1\left(1, 2; 1 - M_2; z_T + \dfrac{S_2 M_2}{2\left(\Gamma(M_2 + 1/2)/\Gamma(1/2)\right)^{M_2}}\right)
$$

$$
(6.75)
$$

and

$$F_{\lambda_1}(\lambda) = \left(\lambda + \frac{S_1 M_1}{2\left(\Gamma(M_1+1/2)/\Gamma(1/2)\right)^{M_1}}\right) \frac{\Gamma(M_1+1)}{(M_1-1)!}$$

$$\times \, _2F_1\left(1,2;1-M_1;\lambda + \frac{S_1 M_1}{2\left(\Gamma(M_1+1/2)/\Gamma(1/2)\right)^{M_1}}\right);$$

$$F_{\lambda_2}(\lambda) = \left(\lambda + \frac{S_2 M_2}{2\left(\Gamma(M_2+1/2)/\Gamma(1/2)\right)^{M_2}}\right) \frac{\Gamma(M_2+1)}{(M_2-1)!}$$

$$\times \, _2F_1\left(1,2;1-M_2;\lambda + \frac{S_2 M_2}{2\left(\Gamma(M_2+1/2)/\Gamma(1/2)\right)^{M_2}}\right). \quad (6.76)$$

6.4.2 SSC Diversity Reception with Correlated Branches

Due to insufficient antenna spacing, both desired and interfering signal envelopes experience correlated Rayleigh fading with joint PDFs, respectively:

$$f_{R_1,R_2}(R_1,R_2) = \left(1-\sqrt{\rho_d}\right) \sum_{k_1,k_2=0}^{\infty} \frac{4\Gamma(k_1+k_2+1)\rho_d^{(k_1+k_2)/2}}{\Gamma(k_1+1)\Gamma(k_2+1)k_1!\,k_2!}$$

$$\times \left(\frac{1}{1+\sqrt{\rho_d}}\right)^{k_1+k_2+1} \left(\frac{1}{\Omega_d\left(1-\sqrt{\rho_d}\right)}\right)^{k_1+k_2+2}$$

$$\times R_1^{2k_1+1} R_2^{2k_2+1} \exp\left(-\frac{R_1^2+R_2^2}{\Omega_d\left(1-\sqrt{\rho_d}\right)}\right) \quad (6.77)$$

and

$$
f_{r_1, r_2}(r_1, r_2) = \frac{\left(1 - \sqrt{\rho_c}\right)^M}{\Gamma(M)} \sum_{l_1, l_2 = 0}^{\infty} \frac{2^2 \Gamma(l_1 + l_2 + 1) \rho_c^{(l_1 + l_2)/2}}{\Gamma(l_1 + 1) \Gamma(l_2 + 1) l_1! l_2!}
$$

$$
\times \left(\frac{1}{1 + \sqrt{\rho_c}}\right)^{l_1 + l_2 + M} \left(\frac{M}{\Omega_c \left(1 - \sqrt{\rho_c}\right)}\right)^{l_1 + l_2 + 2M}
$$

$$
\times e^{-M\left(r_1^2 + r_2^2\right)/\Omega_c \left(1 - \sqrt{\rho_c}\right)}
$$

$$
\times r_1^{2M + 2l_1 - 1} r_2^{2M + 2l_2 - 1}. \tag{6.78}
$$

Therefore, the joint PDF of instantaneous SIRs, denoted by $\lambda_1 = R_1^2/r_1^2$ and $\lambda_2 = R_2^2/r_2^2$, at two input branches can be found by using (6.8) as

$$
f_{\lambda_1 \lambda_2}(\lambda_1, \lambda_2) = \sum_{k_1, k_2, l_1, l_2 = 0}^{\infty} G_{27} \left(S_i \frac{1 - \sqrt{\rho_d}}{1 - \sqrt{\rho_c}}\right)^{2M + l_1 + l_2}
$$

$$
\times \sum_{i=1}^{2} \frac{\lambda_i^{k_i}}{\left(\lambda_i \left(1 - \sqrt{\rho_c}\right) + S_i \left(1 - \sqrt{\rho_d}\right)\right)^{M + k_i + l_i + 1}};
$$

$$
G_{27} = \left(1 - \sqrt{\rho_d}\right)\left(1 - \sqrt{\rho_c}\right)^M \frac{\Gamma(k_1 + k_2 + 1)\Gamma(l_1 + l_2 + M)}{\Gamma(M_i)\Gamma(k_1 + 1)\Gamma(k_2 + 1)}
$$

$$
\times \left(\frac{1}{1 + \sqrt{\rho_d}}\right)^{k_1 + k_2 + 1} \left(\frac{1}{1 + \sqrt{\rho_c}}\right)^{l_1 + l_2 + M}
$$

$$
\times \frac{\Gamma(M + k_1 + l_1 + 1)\Gamma(M + k_2 + l_2 + 1)}{\Gamma(l_1 + M)\Gamma(l_2 + M)k_1! k_2! l_1! l_2!} \rho_d^{(k_1 + k_2)/2} \rho_c^{(l_1 + l_2)/2}, \tag{6.79}
$$

with S_i denoting average input SIRs. Capitalizing on the previous expression with respect to (6.11) and (6.12), it can be obtained:

$$v_{SSC}(\lambda) = \int_{0}^{z_T} f_{\lambda_1\lambda_2}(\lambda,\lambda_2)d\lambda_2 =$$

$$\times \sum_{k_1,k_2,l_1,l_2=0}^{\infty} G_{27} \times \left(S\frac{1-\sqrt{\rho_d}}{1-\sqrt{\rho_c}}\right)^{M+l_1} \frac{\lambda^{k_1}}{\left(\lambda + S\frac{\left(1-\sqrt{\rho_d}\right)}{\left(1-\sqrt{\rho_c}\right)}\right)^{M+k_1+l_1+1}}$$

$$\times B\left(\frac{z_T}{z_T + S\frac{\left(1-\sqrt{\rho_d}\right)}{\left(1-\sqrt{\rho_c}\right)}}, k_2+1, M+l_2\right). \tag{6.80}$$

The CDF of instantaneous SIR at the output of SSC can be expressed according to (6.4), where $F_{\lambda_1\lambda_2}(\lambda, z_T)$ is given by

$$F_{\lambda_1\lambda_2}(\lambda, z_T) = \int_{0}^{\lambda}\int_{0}^{z_T} f_{\lambda_1\lambda_2}(\lambda_1,\lambda_2)d\lambda_1 d\lambda_2$$

$$= \sum_{k_1,k_2,l_1,l_2=0}^{\infty} G_{26} \times B\left(\frac{\lambda}{\lambda + S\frac{\left(1-\sqrt{\rho_d}\right)}{\left(1-\sqrt{\rho_c}\right)}}, k_1+1, M+l_1\right)$$

$$\times B\left(\frac{z_T}{z_T + S\frac{\left(1-\sqrt{\rho_d}\right)}{\left(1-\sqrt{\rho_c}\right)}}, k_2+1, M+l_2\right). \tag{6.81}$$

Figure 6.38 shows the OP versus the normalized switching threshold z_T/S for different number of present interferers. It is obvious that there is an optimal threshold that minimizes OP. For picked normalized

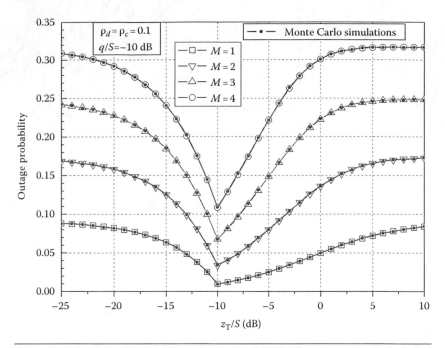

Figure 6.38 OP versus normalized switching threshold in environment with $M=1$, 2, 3, 4 interferers.

outage threshold $q/S=-10$ dB, the normalized switching threshold is also $z_T/S=-10$ dB. So, for this threshold value, SSC combining can be observed as SC. This figure also shows that the higher number of existing interferers degrades the performance gain.

The OP, as a function of normalized parameter S/q for various correlation coefficient values of desired and interfering signals, is shown in Figure 6.39. We observe the SSC receiver as SC one ($q=z_T$). When the correlation coefficient of desired signal ρ_d increases, the OP also increases, as expected. The same influence of coefficient ρ_d on the OP is in environment with one, two, and three interferers. It is interesting to note that when ρ_c increases (the correlation between interferers becomes weaker) the OP improves as the number of interferers increases.

In order to verify analytical results, Monte Carlo simulations were performed and for comparison purposes were included in Figures 6.38 and 6.39. It is evident that analytical results coincide with simulation ones. The samples of correlated Rayleigh fading envelopes were generated by using the generation algorithm presented in [31]. The OP values are estimated on the basis of 10^8 generated samples.

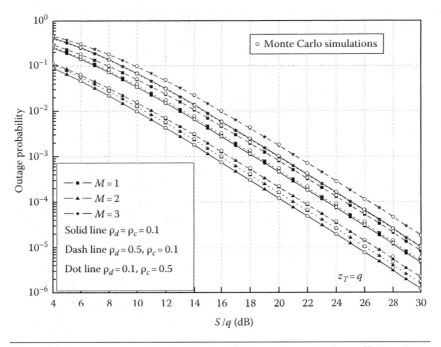

Figure 6.39　OP versus normalized average input SIR for various correlation coefficient values.

6.4.3 SC Diversity Reception with Uncorrelated Branches

Let $\lambda_i = x_i/y_i$ be the SIR at the ith ($i = 1, 2, \ldots, N$) diversity branch of the SC receiver. The joint CDF of multibranch SIR-based SC output could be derived by equating the arguments $\lambda_1 = \lambda_2 = \cdots = \lambda_N$ (since branches are not correlated) as

$$F_{\lambda_i}\left(\lambda_i\right) = \int_0^{\lambda_i} f_{\lambda_i}\left(t_i\right) dt_i$$

$$F_\lambda\left(\lambda\right) = F_{\lambda_1}\left(\lambda\right) F_{\lambda_2}\left(\lambda\right) \cdots F_{\lambda_N}\left(\lambda\right) = \prod_{i=1}^N F_{\lambda_i}\left(\lambda\right);$$

$$F_\lambda\left(\lambda\right) = \prod_{i=1}^N \left(\lambda + \frac{S_i M_i}{2\left(\Gamma\left(M_i + 1/2\right)/\Gamma\left(1/2\right)\right)^{M_i}}\right) \frac{\Gamma\left(M_i + 1\right)}{\left(M_i - 1\right)!}$$

$$\times {}_2F_1\left(1, 2; 1 - M_i; \lambda + \frac{S_i M_i}{2\left(\Gamma\left(M_i + 1/2\right)/\Gamma\left(1/2\right)\right)^{M_i}}\right), \quad (6.82)$$

with $S_i = \Omega_{xi}/\Omega_{0yi}$ being the average SIRs at the ith input branch of the selection combiner system.

The PDF at the output of the SC can be found as

$$f_\lambda(\lambda) = \frac{d}{d\lambda}F(\lambda) = \sum_{i=1}^{n} f_{\lambda_i}(\lambda) \prod_{\substack{j=1 \\ j \neq i}}^{n} F_{\lambda_i}(\lambda); \quad f_{\lambda_i}(\lambda) = \frac{d}{d\lambda}F_{\lambda_i}(\lambda);$$

$$f_\lambda(\lambda) = \sum_{i=1}^{N} \frac{\left(\dfrac{S_i M_i}{2\left(\Gamma(M_i+1/2)/\Gamma(1/2)\right)^{M_i}}\right)^{M_i} \Gamma(M_i+1)}{(M_i-1)!\left(\lambda + \dfrac{S_i M_i}{2\left(\Gamma(M_i+1/2)/\Gamma(1/2)\right)^{M_i}}\right)^{M_i+1}}$$

$$\times \prod_{\substack{j=1 \\ j \neq i}}^{N} \left(\lambda + \frac{S_j M_j}{2\left(\Gamma(M_j+1/2)/\Gamma(1/2)\right)^{M_j}}\right) \frac{\Gamma(M_j+1)}{(M_j-1)!}$$

$$\times \, _2F_1\left(1,2;1-M_j;\lambda + \frac{S_j M_j}{2\left(\Gamma(M_j+1/2)/\Gamma(1/2)\right)^{M_j}}\right). \quad (6.83)$$

OP for the uncorrelated case versus normalized parameter S_1/γ for balanced and unbalanced ratio of SIR at the input of the branches and various values of the number of multiple interferers diversity order is shown in Figures 6.40 and 6.41. From Figure 6.40, we can see the way OP increases when the number of multiple independent CCIs increase due to growth of interference domination. For example for the dual-branch case, the presence of two interferers instead of one degrades the performances more than about 2.5 dB, the presence of three interferers degrades another 2 dB, and the presence of four interferers degrades about 1 dB more. In other words, considering four interferers instead of one decreases performances approximately 5.5 dB in the wide range of requested OPs (OP = 10^{-1}–10^{-4}). Also, for this dual-SC diversity case, we can see how OP deteriorates when input SIR unbalance is present. From Figure 6.41, we can see that for the constant number of CCIs OP behavior improves as the diversity order (number of branches) increases. For example, considering the

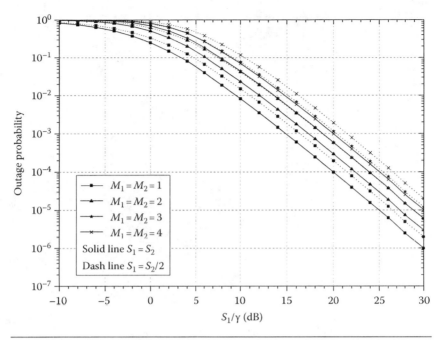

Figure 6.40   OP for the dual-branch SC with uncorrelated case for balanced and unbalanced ratios of SIR at the input branches and various values of the number of multiple interferers.

case of two interferers we can see that at the OP = 10^{-1} improvement offered by triple SC is about 1.5 dB and by four-branch SC diversity receiver another 2.5 dB. For OP = 10^{-2} the corresponding improvements are 2 dB and 4 dB, respectively. The gains offered at the OP = 10^{-3} by the usage of triple- and four-branch diversity receiver are 2.6 and 5 dB, respectively. Finally, at OP = 10^{-4} the corresponding improvements are 3.2 and 7 dB, respectively. Similar conclusions could be drawn for the input SIR unbalance at the diversity branches. OP deteriorates when the input SIR unbalance is present and the degradation increases with the number of branches.

6.4.4 SC Diversity Reception with Correlated Branches

Let us assume that there are M correlated Rayleigh distributed interferers over kth branch of the SC diversity system with N branches, having the same average power. We will discuss the following case considering the proposed model of constant correlation between the branches for the Nakagami-m model, given in [17]. It is a

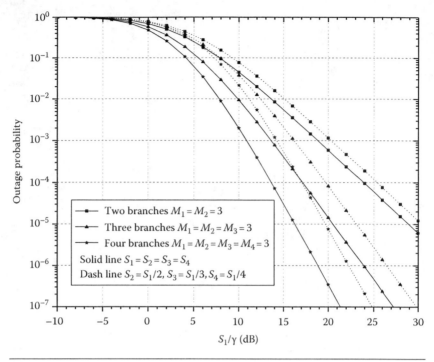

Figure 6.41 OP for balanced and unbalanced ratios of SIR at the input of the branches and various values of diversity order.

well-known fact that one variable Nakagami-m (and Rayleigh) distribution could be derived from the central χ^2 (chi-square) distribution with $2L$ degrees of freedom (L independent complex Gaussian RVs) [32]. Knowing the definition of Nakagami-m fading average power and severity parameter and the fact that Rayleigh distribution can be obtained from Nakagami-m as the special case for $m=1$, it can be easily shown that the following relation between the powers stands $\Omega_k/L = \Omega_k/m = 2\sigma^2$ [32]. Let ρ_d be the correlation coefficient for the desired signal, defined as $cov(R_i^2, R_j^2)/(var(R_i^2)var(R_j^2))^{1/2}$, while the correlation coefficient ρ_c for the interfering signal is defined as $cov(r_i^2, r_j^2)/(var(r_i^2)var(r_j^2))^{1/2}$, with R_k and r_k denoting the amplitudes of the desired and interference signals received at the kth branch. The constant correlation model [17] can be obtained by setting $\Sigma_{i,j} \equiv 1$ for $i=j$ and $\Sigma_{i,j} \equiv \rho$ for $i \neq j$ in the correlation matrix, for both desired signal and interference. Now joint distributions of PDF for both desired and interfering signal correlated envelopes for a multibranch signal combiner could be expressed by

$$f_{R_1,\ldots,R_N}\left(R_1,\ldots,R_N\right)$$

$$= (1-\sqrt{\rho_d})\underbrace{\sum_{k_1=0}^{\infty}\cdots\sum_{k_n=0}^{\infty}}_{n}\left[\frac{2^n}{\Gamma(1+k_1)\cdots\Gamma(1+k_n)}\frac{\Gamma(1+k_1+\cdots+k_n)}{k_1!\cdots k_n!}\right.$$

$$\times\rho_d^{(k_1+\cdots+k_n)/2}\left(\frac{1}{1+(n-1)\sqrt{\rho_d}}\right)^{1+k_1+\cdots+k_n}\left(\frac{1}{\Omega_{d1}(1-\sqrt{\rho_d})}\right)^{1+k_1}\cdots$$

$$\times\left(\frac{1}{\Omega_{dn}(1-\sqrt{\rho_d})}\right)^{1+k_n}R_1^{2k_1+1}\cdots R_n^{2k_n+1}\exp\left(\frac{R_1^2}{\Omega_{d1}(1-\sqrt{\rho_d})}\right)\cdots$$

$$\times\exp\left(\frac{R_n^2}{\Omega_{dn}(1-\sqrt{\rho_d})}\right)\Bigg]; \tag{6.84}$$

$$f_{r_1,\ldots,r_N}\left(r_1,\ldots,r_N\right)$$

$$= \frac{(1-\sqrt{\rho_c})^M}{\Gamma(M)}\underbrace{\sum_{l_1=0}^{\infty}\cdots\sum_{l_n=0}^{\infty}}_{n}\left[\frac{2^n}{\Gamma(M+l_1)\cdots\Gamma(M+l_n)}\frac{\Gamma(M+l_1+\cdots+l_n)}{l_1!\cdots l_n!}\right.$$

$$\times\rho_c^{\frac{l_1+\cdots+l_n}{2}}\left(\frac{1}{1+(n-1)\sqrt{\rho_c}}\right)^{1+l_1+\cdots+l_n}\left(\frac{M}{\Omega_{c1}(1-\sqrt{\rho_d})}\right)^{M+l_1}\cdots$$

$$\times\left(\frac{M}{\Omega_{cn}(1-\sqrt{\rho_d})}\right)^{M+l_n}r_1^{2M+2l_1-1}\cdots r_n^{2M+2l_n-1}\exp\left(\frac{Mr_1^2}{\Omega_{c1}(1-\sqrt{\rho_c})}\right)\cdots$$

$$\times\exp\left(\frac{Mr_n^2}{\Omega_{cn}(1-\sqrt{\rho_c})}\right)\Bigg]; \tag{6.85}$$

$\Omega_{dk}=\overline{R_k^2}$ stands for the average desired signal power at kth branch

$\Omega_{ck}=\overline{r_k^2}$ is the total average interference signal power at kth branch, where Ω_{cik}, $i=1\ldots M$, is the average power of the single CCI and stands $\Omega_{ck}=M\,\Omega_{cik}$, while M denotes the number of interferers over kth branch.

Let us consider the instantaneous value of SIR at the kth diversity branch input in the form of $\lambda_k = R_k^2/r_k^2$. The joint PDF of instantaneous values of SIR at the input of a multibranch SC combiner could be obtained as in (6.30), while the joint CDF can be written as (6.33). Let $S_k = \Omega_{dk}/\Omega_{cik} = \Omega_{dk}/(\Omega_{ck}/M)$ be the average SIRs at the kth input branch of the multibranch SC. The CDF of output SIR could be derived by equating the arguments $t_1 = \cdots = t_n = t$ as in (6.35):

$$F_\lambda(t)$$

$$= \sum_{\substack{k_1,\ldots,k_n=0 \\ 2n}}^{\infty} \sum_{l_1,\ldots,l_n=0}^{\infty} G_{28} t^{n+k_1+\cdots+k_n} \frac{\displaystyle\prod_{i=1}^{n} {}_2F_1\left(\begin{array}{c} 1+k_i, 1-M-l_i; \\[2mm] 2+k_i; \dfrac{t}{t+\dfrac{(1-\sqrt{\rho_d})}{(1-\sqrt{\rho_c})}S_i} \end{array} \right)}{\left(t+\dfrac{(1-\sqrt{\rho_d})}{(1-\sqrt{\rho_c})}S_i \right)^{1+k_i}};$$

$$G_{28} = \frac{\begin{array}{c} (1-\sqrt{\rho_d})(1-\sqrt{\rho_c})^M \Gamma(1+k_1+\cdots+k_n)\Gamma(M+l_1+\cdots+l_n) \\[2mm] \times \Gamma(1+M+k_1+l_1)\cdots\Gamma(1+M+k_n+l_n) \end{array}}{\Gamma(M)(1+k_1)\cdots(1+k_n)\Gamma(1+k_1)\cdots\Gamma(1+k_n)}$$

$$\times \Gamma(M+l_1)\cdots\Gamma(M+l_n)k_1!\cdots k_n!l_1!\cdots l_n!$$

$$\times \rho_d^{\frac{k_1+\cdots+k_n}{2}} \rho_c^{\frac{l_1+\cdots+l_n}{2}} \left(\frac{1}{1+(n-1)\sqrt{\rho_d}} \right)^{1+k_1+\cdots+k_n}$$

$$\times \left(\frac{1}{1+(n-1)\sqrt{\rho_c}} \right)^{1+l_1+\cdots+l_n}. \tag{6.86}$$

In Table 6.5, the number of terms to be summed in order to achieve accuracy at the desired significant digit is depicted. As we can see from the table, the values of these terms are strongly related to the correlation coefficients ρ_d and ρ_c and the number of interferers M.

Now, the PDF of the SC output SIR can be obtained easily from the previous expression:

Table 6.5 Terms That Need to Be Summed in Expression (6.86) for CDF of Triple-Branch SC Output SIR Case to Achieve Accuracy at the 4th Significant Digit Presented in the Brackets

$M=2$, $\rho_d=\rho_c=0.2$, $S_1=S_2$		$M=3$, $\rho_d=\rho_c=0.2$, $S_1=S_2$	
$S/\lambda=-10$ dB	$S/\lambda=0$ dB	$S/\lambda=-10$ dB	$S/\lambda=0$ dB
15	16	17	17
19	19	22	20
22	21	24	22

$$f_\lambda(t) = \frac{d}{dt} F_\lambda(t)$$

$$= \underbrace{\sum_{k_1,\ldots,k_n=0}^{\infty} \sum_{l_1,\ldots,l_n=0}^{\infty}}_{2n} G_{28}(1+k_1)\cdots(1+k_n)t^{n+k_1+\cdots+k_n}(A_1(t)+\cdots+A_n(t))$$

$$A_i(t) = \left(\frac{S_i}{t+\frac{(1-\sqrt{\rho_d})}{(1-\sqrt{\rho_c})}S_i}\right)^{M+l_i} \prod_{\substack{j=1 \\ j\neq i}}^{n} \frac{{}_2F_1\left(\begin{array}{c}1+k_j, 1-M-l_j; \\ 2+k_j; \dfrac{t}{t+\dfrac{(1-\sqrt{\rho_d})}{(1-\sqrt{\rho_c})}S_j}\end{array}\right)}{(1+k_j)}.$$

$$(6.87)$$

Figures 6.42 and 6.43 show the OP for the correlated case versus normalized parameter S_1/γ for balanced and unbalanced ratios of SIR at the input of the branches, various values of the number of multiple interferers, and the level of correlation. It can be seen from Figure 6.42 how the OP increases when the number of multiple CCIs increases due to the growth of interference domination. Also for this dual-SC diversity case, it is evident how the OP deteriorates at a higher level of correlation between the diversity branches. Considering the case when the correlation levels between the desired signal envelopes and between the interferers ρ_d and ρ_c arise from $\rho_d=\rho_c=0.2$ to $\rho_d=\rho_c=0.4$, we can observe that performances decrease in approximately 0.5 dB over the wide range of requested OP (OP $= 10^{-1}-10^{-4}$). Figure 6.43

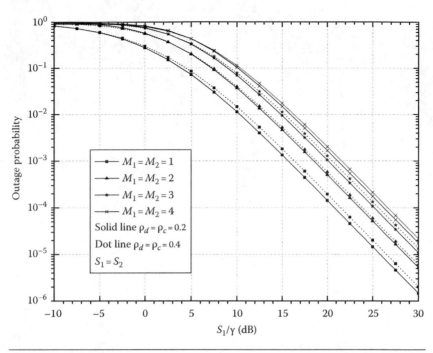

Figure 6.42 OP for the two-branch correlated case versus normalized parameter S_1/γ for various values of the number of multiple interferers and the level of correlation.

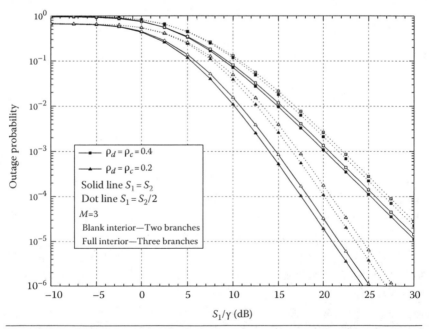

Figure 6.43 OP for the correlated case for balanced and unbalanced ratios of SIR at the input of the branches, various values of level of correlation, and diversity order.

also shows an improvement of the OP behavior for the constant number of CCIs and with the increased level of correlation. It can be observed how the OP behavior improves as the diversity order (number of branches) increases. For example, considering the case of the presence of $M = 3$ interferers, we can see that at $OP = 10^{-1}$ improvement offered by triple SC is about 3.5 dB. For $OP = 10^{-2}$ the corresponding improvement is about 5 dB. The gains at $OP = 10^{-3}$ is about 7 dB. Finally, for $OP = 10^{-4}$ the corresponding improvement is about 8.2 dB. Similar conclusions could be drawn for the input SIR unbalance at the diversity branches. The OP deteriorates when the input SIR unbalance is present, and the degradation increases with the number of branches. Similar conclusions could be drawn for the input SIR balance effect between the diversity branches. Based on previous evaluation of the presented numerical results and presented performance trade-offs, by using derived expressions, engineering decisions could be made to determine optimal values of system parameters for achieving a reasonable level of outage in practical wireless applications.

References

1. Simon, M. K. and Alouini, M. S. (2005). *Digital Communications over Fading Channels*, 2nd edn. Wiley, New York.
2. Ko, Y. C., Alouni, M. S., and Simon, M. K. (2000). Analysis and optimization of switched diversity systems. *IEEE Transactions on Communications*, 49(5), 1813–1831.
3. Bithas, P. S., Mathiopoulos, P. T., and. Karagiannidis, G. K. (2006). Switched diversity receivers over correlated Weibull fading channels. *Proceedings of the International Workshop on Satellite and Space Communications*, Madrid, Spain, pp. 143–147, September 2006.
4. Gradshteyn, I. and Ryzhik, I. (1980). *Tables of Integrals, Series, and Products*. Academic Press, New York.
5. Yacoub, M. D. (2007). The α-μ distribution: A physical fading model for the Stacy distribution. *IEEE Transactions on Vehicular Technology*, 56(1), 27–34.
6. Wolfram Research, Inc. http://mathworld.wolfram.com/Laguerre Polynomial.html, accessed January 2009.
7. Helstrom, C. W. (1991). *Probability and Stochastic Processes for Engineers*, 2nd edn. MacMillan, London, U.K.
8. Spalevic, P. and Panić, S. (2010). SSC diversity receiver over correlated α-μ fading channels in the presence of co-channel interference. *EURASIP Journal on Wireless Communications and Networking*, 2010, Article ID 142392, 1–7.

9. Panic, S. et al. (2011). On the performance analysis of SIR-based SSC diversity over correlated α-μ fading channels. *Computers and Electrical Engineering*, 37(3), 332–338.

10. Karagiannidis, G. K. (2003). Performance analysis of SIR-based dual selection diversity over correlated Nakagami-m fading channels. *IEEE Transactions on Vehicular Technology*, 52(5), 1207–1216.

11. Hamdi, K. A. (2008). Moments and autocorrelations of the signal to interference ratio in wireless communications. *Proceedings of the International Conference on Communications*, Beijing, China, pp. 1345–1348.

12. Aalo, V. A. (1995). Performance of maximal-ratio diversity systems in a correlated Nakagami fading environment. *IEEE Transactions on Communications*, 43(8), 2360–2369.

13. Reig, J. (2007). Multivariate Nakagami-*m* distribution with constant correlation model. *International Journal of Electronics and Communications (AEUE)*, 63(1), 46–51.

14. Yacoub, M. D. (2002). The α-μ distribution: A general fading distribution. *Proceedings of the 13th International Symposium on Personal, Indoor, and Mobile Radio Communications*, Lisaboa, Portugal, Vol. 2, 629–633.

15. Proakis, J. G. (2001). *Digital Communications*. McGraw-Hill, New York, 2001.

16. Helstrom, C. W. (1984). *Probability and Stochastic Processes for Engineers*. MacMillan, New York.

17. Reig, J. et al. (2007). Generation of bivariate Nakagami-m fading envelopes with arbitrary not necessary identical fading parameters. *Wireless Commutations and Mobile Computing*, 7, 531–537.

18. Marsaglia, G. and Tsang, W. W. (2000). A simple method for generating gamma variables. *ACM Transactions on Mathematical Software*, 26(3), 363–372.

19. Lee, W. C. Y. (1993). *Mobile Communications Design Fundamentals*. Wiley, New York.

20. Karagiannidis, G. K., Zogas, D. A., and Kotsopoulos, S. A. (2003). On the multivariate Nakagami-m distribution with exponential correlation. *IEEE Transactions on Communications*, 51(8), 1240–1244.

21. De Souza, R. A. A., Fraidenraich, G., and Yacoub, M. D. (2006). On the multivariate α-μ distribution with arbitrary correlation. *Proceedings of VI International Telecommunications Symposium*, Fortaleza, Ceara, Brasil, pp. 38–41.

22. Stefanovic, M. et al. (2008). Performance analysis of system with selection combining over correlated Weibull fading channels in the presence of co-channel interference. *International Journal of Electronics and Communications (AEUE)*, 62(9), 695–700.

23. Bandjur, D. V., Stefanovic, M. C., and Bandjur, M. V. (2008). Performance analysis of SSC diversity receivers over correlated Ricean fading channels in the presence of co-channel interference. *Electronic Letters*, 44(9), 587–588.

24. Nikolic, B. et al. (2011). Selection combining system over correlated generalized-K (K$_G$) fading channels in the presence of co-channel interference. *ETRI Journal*, 33(3), 320–325.
25. Bithas, P. S., Sagias, N. C., and Mathiopoulos, P. T. (2009). The bivariate generalized-K (K$_G$) distribution and its application to diversity receivers. *IEEE Transactions on Communications*, 57(9), 2655–2662.
26. Trigui, I. et al. (2009). Performance analysis of mobile radio systems over composite fading/shadowing channels with co-located interferences. *IEEE Transactions on Wireless Communications*, 8(7), 3449–3453.
27. Trigui, I. et al. (2009). Outage analysis of wireless systems over composite fading/shadowing channels with co-channel interferences. *Proceedings of IEEE Wireless Communications and Networking Conference*, Budapest, Hungary, pp. 1–6.
28. Alouini, M. and Goldsmith, A. (1999). Area spectral efficiency of cellular mobile radio systems. *IEEE Transactions on Vehicular Technology*, 48(4), 1047–1065.
29. Kang, M., Alouini, M., and Yang, L. (2002). Outage probability and spectral efficiency of cellular mobile radio systems with smart antennas. *IEEE Transactions on Vehicular Technology*, 50(12), 1871–1877.
30. Bhaskar, V. (2009). Capacity evaluation for equal gain diversity schemes over Rayleigh fading channels. *International Journal of Electronics and Communications (AEUE)*, 63(4), 235–240.
31. Ertel, R. and Reed, J. (1998). Generation of two equal power correlated Rayleigh fading envelopes. *IEEE Communications Letters*, 2(10), 276–278.
32. Luo, J. and Zelder, J. (2000). A statistical simulation model for correlated Nakagami fading channels. *Proceedings of the International Conference on Communications Technology*, Beijing, China, Vol. 2, pp. 1680–1684.

7

MACRODIVERSITY RECEPTION OVER FADING CHANNELS IN THE PRESENCE OF SHADOWING

Wireless channels are simultaneously affected by short-term fading and long-term fading (shadowing) [1]. The short-term signal variation has been described by several distributions such as Hoyt, Rayleigh, Rice, Nakagami-m, κ-μ, η-μ, α-μ, and Weibull (see Chapter 2). Various techniques for reducing short-term fading effect are used in wireless communication systems [2]. An efficient method for amelioration system's quality of service (QoS) by using multiple receiver antennas is called space diversity. The main goal of space diversity techniques is to upgrade transmission reliability without increasing transmission power and bandwidth while increasing channel capacity. Several principal types of combining techniques can be generally performed by their dependence on complex restrictions put on the communication system and the amount of channel-state information available at the receiver. In order to implement combining techniques such as maximum ratio combining (MRC) and equal gain combining (EGC), it is not only necessary to know all or some of the channel state information (CSI) of the received signal, but also the receiver chain for each branch of the diversity system. It increases the complexity of the system. Unlike previous combining techniques, a selection combining (SC) receiver processes only one of the diversity branches and is much simpler and cheaper for practical realization (see Chapter 4).

While short-term fading is mitigated through the use of diversity techniques typically at the single base station (microdiversity), use of only such microdiversity approaches will not be sufficient to mitigate the overall channel degradation when shadowing is also concurrently present. In cellular networks, long-term fading known as shadowing can put a heavy limitation on the system's performance. Shadowing is the result of the topographical elements and other structures in the transmission path such as trees and tall buildings. Now, we must simultaneously consider both short- and long-term fading conditions since they coexist in wireless systems [3]. Macrodiversity is used to alleviate the effects of shadowing where multiple signals are received at widely located base stations (BSs), ensuring that different long-term fading is experienced by these signals [4]. The simultaneous use of multiple base stations and the processing of signals from these base stations will provide the framework for both macro- and microdiversity techniques to improve the performance in shadowed fading channels [5–9].

In this chapter, we will discuss the necessity and the validity of macrodiversity reception usage, from the point of view of multipath fading and shadowing mitigation, through the second-order statistical measures at the output of the macrodiversity receiver. Most of the results in this chapter and detailed analysis can be found in the literature reported by the contributing authors [10,11].

7.1 SC Macrodiversity System Operating over Gamma-Shadowed Nakagami-*m* Fading Channels

In this chapter, MRC combining at microlevel (at each base station) will be considered, while SC between BSs (macrolevel) is applied. At the macrolevel, SC will be observed, as basically a fast response hand-off of mechanism that instantaneously or with minimal delay chooses the best BS to serve mobile devices based on the signal power received [12]. The observed selection will be based on the received signal power. We will consider cases when macrodiversity inputs are both correlated and uncorrelated, since the level of correlation depends on the separation between the BSs, the surrounding terrain, the angle of arrival of the received signals, and various other factors. Analysis will be carried out through the second-order statistical measures at the output of the macrodiversity receiver, such as level crossing rate (LCR) and average fade duration (AFD). Since the Nakagami-*m* fading model provides a good fit to the collected data in indoor and outdoor mobile-radio environments, it will be used for modeling signal propagation and received statistics at microdiversity inputs, while the received signal powers will be modeled by gamma distributions.

7.1.1 Uncorrelated Shadowing

First we will observe a case when BSs at the macrodiversity level are widely located, due to sufficient spacing between antennas, so signal powers at the outputs of the base stations will be modeled as statistically independent. A macrodiversity system is of SC type and consists of two microdiversity systems with selection based on their output signal power values. Each microdiversity system is of MRC type with an arbitrary number of branches in the presence of Nakagami-m fading. Treating the correlation between the branches at microlevel as exponential, the expression for the probability density function (PDF) of the SNR at the outputs of microdiversity systems is as follows [13]:

$$f_{z_i/\Omega_i}\left(\frac{z_i}{\Omega_i}\right) = \frac{1}{\Gamma(M_i)}\left(\frac{N_i m_i}{r_i \Omega_i}\right)^{M_i} z_i^{M_i-1} \exp\left(-\frac{N_i m_i}{r_i \Omega_i} z\right). \quad (7.1)$$

In the previous equation, $\Gamma(x)$ denotes the gamma function [9, Eq. (8.310 7.1)], m_i is the well-known Nakagami-m fading severity parameter, N_i denotes the number of identically assumed channels at each microlevel, while parameters r_i, related to the exponential correlation ρ_i among the branches, and parameter M_i are given with

$$r_i = N_i + \frac{2\rho_i}{1-\rho_i}\left[N_i - \frac{1-\rho_i^{N_i}}{1-\rho_i}\right]; \quad M_i = \frac{m_i N_i^2}{r_i}. \quad (7.2)$$

Since the outputs of MRCs and their time derivatives [14] are

$$z_i^2 = \sum_{k=1}^{N_i} z_{ik}^2; \quad \dot{z}_i = \sum_{k=1}^{N_i} \frac{z_{ik}}{z_i}\dot{z}_{ik} \quad i=1,2; \quad (7.3)$$

$\dot{z}_i$ is a Gaussian random variable with zero mean

$$f(\dot{z}_i) = \frac{1}{\sqrt{2\pi}\dot{\sigma}_{z_i}} \cdot \exp\left(-\frac{\dot{z}_i^2}{2\dot{\sigma}_{z_i}^2}\right); \quad (7.4)$$

and variance given by [15]

$$\dot{\sigma}_{z_i}^2 = \sum_{k=1}^{N_i} \frac{z_{ik}^2 \dot{\sigma}_{z_{ik}}^2}{z_i^2}. \tag{7.5}$$

For the case of equivalently assumed channels, when $\dot{\sigma}_{z_{i1}}^2 = \dot{\sigma}_{z_{i2}}^2 = \cdots = \dot{\sigma}_{z_{ik}}^2$; $k = 1,\ldots,N$, the previous equation reduces to

$$\dot{\sigma}_{z_i}^2 = \dot{\sigma}_{z_{ik}}^2 = \frac{\dot{\pi} f_d^2 \, \Omega_i}{m_i}, \tag{7.6}$$

where f_d denotes a Doppler shift frequency. Conditioned on Ω_i, the JPDF $f(z_i, \dot{z}_i/\Omega_i)$ can be presented as

$$f_{z_i,\dot{z}_i/\Omega_i}\left(z_i, \frac{\dot{z}_i}{\Omega_i}\right) = f_{z_i/\dot{z}_i,\Omega_i}\left(\frac{z_i}{\dot{z}_i}, \Omega_i\right) \times f_{\dot{z}_i/\Omega_i}\left(\frac{\dot{z}_i}{\Omega_i}\right) =$$

$$= \frac{z_i^{M_i-1}}{\Gamma(M_i)}\left(\frac{N_i m_i}{r_i \Omega_i}\right)^{M_i} \exp\left(-\frac{N_i m_i}{r_i \Omega_i} z_i\right)$$

$$\times \frac{1}{\sqrt{2\pi}\dot{\sigma}_{z_i}} \exp\left(-\frac{\dot{z}_i^2}{2\dot{\sigma}_{z_i}^2}\right); \quad i = 1, 2. \tag{7.7}$$

It has already been noted that our macrodiversity system is of SC type and that the selection is based on the microcombiner output signal power values. This selection can be written through the first-order statistical parameters PDF and cumulative distribution function (CDF) at the macrodiversity output in the form of

$$f_{z\dot{z}}(z,\dot{z}) = \int_0^\infty d\Omega_1 \int_0^{\Omega_1} d\Omega_2 f_{z_1\dot{z}_1/\Omega_1}\left(\frac{z,\dot{z}}{\Omega_1}\right) f_{\Omega_1\Omega_2}\left(\Omega_1\Omega_2\right)$$

$$+ \int_0^\infty d\Omega_2 \int_0^{\Omega_2} d\Omega_1 f_{z_2\dot{z}_2/\Omega_1}\left(\frac{z,\dot{z}}{\Omega_2}\right) f_{\Omega_1\Omega_2}\left(\Omega_1\Omega_2\right) \tag{7.8}$$

and

$$F_z(z) = \int_0^\infty d\Omega_1 \int_0^{\Omega_1} d\Omega_2 F_{z_1/\Omega_1}\left(\frac{z}{\Omega_1}\right) f_{\Omega_1\Omega_2}(\Omega_1\Omega_2)$$

$$+ \int_0^\infty d\Omega_2 \int_0^{\Omega_2} d\Omega_1 F_{z_2/\Omega_2}\left(\frac{z}{\Omega_2}\right) f_{\Omega_1\Omega_2}(\Omega_1\Omega_2);$$

$$F\left(\frac{z_i}{\Omega_i}\right) = \int_0^{z_i} f\left(\frac{t_i}{\Omega_i}\right) dt_i. \tag{7.9}$$

Here, long-term fading influence at the uncorrelated microdiversity branches is, as in [16], described with gamma distributions, which are independent (as mentioned earlier):

$$f_{\Omega_1,\Omega_2}(\Omega_1,\Omega_2) = f_{\Omega_1}(\Omega_1) \times f_{\Omega_2}(\Omega_2) = \frac{1}{\Gamma(c_1)} \cdot \frac{\Omega_1^{c_1-1}}{\Omega_{01}^{c_1}} \cdot \exp\left(-\frac{\Omega_1}{\Omega_{01}}\right)$$

$$\times \frac{1}{\Gamma(c_2)} \cdot \frac{\Omega_2^{c_2-1}}{\Omega_{02}^{c_2}} \exp\left(-\frac{\Omega_2}{\Omega_{02}}\right). \tag{7.10}$$

In the previous equation, c_1 and c_2 denote the order of the gamma distribution, the measure of the shadowing present in the channels. The parameters Ω_{01} and Ω_{02} are related to the average powers of the gamma long-term fading distributions.

Let z be the received signal envelope and $\dot{z}$ its derivative with respect to time. The LCR at the envelope z is defined as the rate at which a fading signal envelope crosses level z in a positive or a negative direction and is mathematically defined by formula (4.13). The AFD is defined as the average time over which the signal envelope ratio remains below a specified level after crossing that level in a downward direction and is determined as (4.14). After substituting (7.7), (7.10), and (7.8) into (4.13), and following the procedure explained in Appendix from [10], we can easily derive the infinite-series expression for the system output LCR, in the form of

$$N_Z(z) = \int_0^\infty \dot{z} f_{z\dot{z}}(z,\dot{z}) d\dot{z};$$

$$\frac{N_z(z)}{f_d} = \frac{z^{M_1-1}}{\Gamma(M_1)\Gamma(c_1)\Gamma(c_2)} \left(\frac{N_1 m_1}{r_1}\right)^{M_1}$$

$$\times \sqrt{\frac{2\pi}{m_1}} \sum_{k=0}^{\infty} \frac{\left(N_1 m_1 z / r_1 (1/\Omega_{01} + 1/\Omega_{02})\right)^{(M_1+c_1+c_2+k-1/2)/2}}{c_2(1+c_2)_k \Omega_{01}{}^{c_1} \Omega_{02}{}^{k+c_2}}$$

$$\times K_{(M_1+c_1+c_2+k-1/2)}\left(2\sqrt{\frac{N_1 m_1 z(\Omega_{01}+\Omega_{02})}{r_1 \Omega_{01}\Omega_{02}}}\right)$$

$$+ \frac{z^{M_2-1}}{\Gamma(M_2)\Gamma(c_1)\Gamma(c_2)} \left(\frac{N_2 m_2}{r_2}\right)^{M_2}$$

$$\times \sqrt{\frac{2\pi}{m_2}} \sum_{k=0}^{\infty} \frac{\left(N_2 m_2 z / r_2 (1/\Omega_{01} + 1/\Omega_{02})\right)^{(M_2+c_1+c_2+k-1/2)/2}}{c_2(1+c_2)_k \Omega_{01}{}^{c_1} \Omega_{02}{}^{k+c_2}}$$

$$\times K_{(M_2+c_1+c_2+k-1/2)}\left(2\sqrt{\frac{N_2 m_2 z(\Omega_{01}+\Omega_{02})}{r_2 \Omega_{01}\Omega_{02}}}\right). \tag{7.11}$$

In a similar manner from (4.14), we can obtain an infinite series expression for the output AFD in the form of

$$T_z(z) = \frac{F_z(z \le Z)}{N_z(z)};$$

$$F_z(z) = \frac{2(N_1 m_1 / r_1)^{M_1}}{\Gamma(M_1)\Gamma(c_1)\Gamma(c_2)M_1} \sum_{k=0}^{\infty} \sum_{l=0}^{\infty} \left(\frac{N_1 m_1}{r_1}\right)^k$$

$$\times \frac{\left(N_1 m_1 z / r_1 (1/\Omega_{01} + 1/\Omega_{02})\right)^{(c_1+c_2+l-k-M_1)/2}}{\Omega_{01}{}^{c_1}\Omega_{02}{}^{l+c_2} c_2(1+c_2)_l (1+M_1)_k}$$

$$\times K_{(c_1+c_2+l-k-M_1)}\left(2\sqrt{\frac{N_1 m_1 z(\Omega_{01}+\Omega_{02})}{r_1 \Omega_{01}\Omega_{02}}}\right)$$

$$+ \frac{2(N_2 m_2 / r_2)^{M_2}}{\Gamma(M_2)\Gamma(c_1)\Gamma(c_2)M_2} \sum_{k=0}^{\infty} \sum_{l=0}^{\infty} \left(\frac{N_2 m_2}{r_2}\right)^k$$

$$\times \frac{\left(N_2 m_2 z / r_2 (1/\Omega_{01} + 1/\Omega_{02})\right)^{(c_1+c_2+l-k-M_2)/2}}{\Omega_{01}^{l+c_2} \Omega_{02}^{c_1} c_1 (1+c_1)_l (1+M_2)_k}$$

$$\times K_{(c_1+c_2+l-k-M_2)} \left(2 \sqrt{\frac{N_2 m_2 z (\Omega_{01} + \Omega_{02})}{r_2 \Omega_{01} \Omega_{02}}} \right), \qquad (7.12)$$

where

$K_n(.)$ is the nth-order-modified Bessel function of the second kind [9, Eq. (8.407)]

$(a)_n$ denotes the Pochhammer symbol [9]

In order to show the influence of various parameters such as the number of the diversity branches at the microcombiners, fading severity, and the level of correlation between those branches on the system's statistics, numerical results are given and graphically presented. Normalized values of LCR, by maximal Doppler shift frequency f_d, are presented in Figures 7.1 and 7.2.

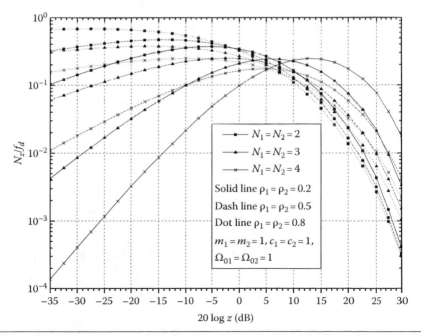

Figure 7.1 Normalized average LCR of our macrodiversity structure for various values of correlation level and diversity order.

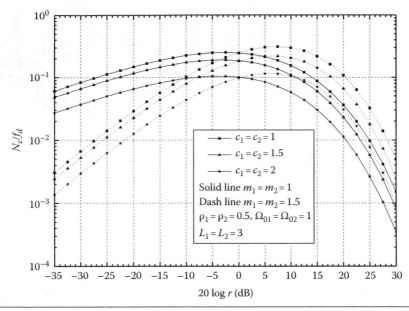

Figure 7.2 Normalized average LCR of our macrodiversity structure for various values of fading and shadowing severity level.

We can observe from Figure 7.1 that lower levels are crossed with the higher number of diversity branches at each microcombiner and lower level of correlation between the branches. From Figure 7.2, it is obvious that larger values of Nakagami-m fading and shadowing severity parameters m_i and c_2 provide smaller LCR values.

The normalized AFD for various values of the system's parameters is presented in Figures 7.3 and 7.4. Similarly, better performances of system are achieved (lower values of AFD) with a higher number of diversity branches, smaller correlation level, and a higher number of fading and shadowing severity.

In Figures 7.5 and 7.6, the normalized AFD and LCR values of our macrodiversity system are compared to AFD and LCR values of non-diversity cases. Considering [6, Eq. (3)] and [17, Eq. (8)], with respect to (7.6) through (7.8) and (4.14), AFD can be efficiently obtained for the case of gamma-shadowed Nakagami-m composite fading channel, when neither macro- nor microdiversity is applied. It is visible from Figure 7.5 that AFD values are lower for the case of simultaneous fading/shadowing cancellation, which we proposed in our work.

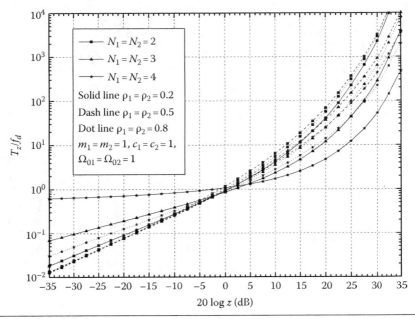

Figure 7.3   Normalized AFD of our macrodiversity structure for various values of correlation level and diversity order.

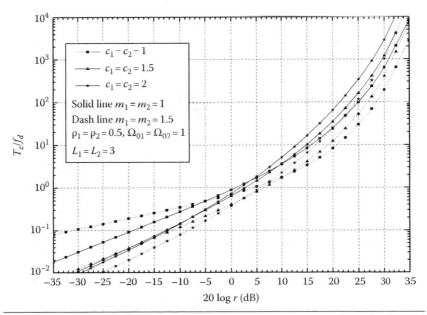

Figure 7.4   Normalized AFD of our macrodiversity structure for various values of fading and shadowing severity level.

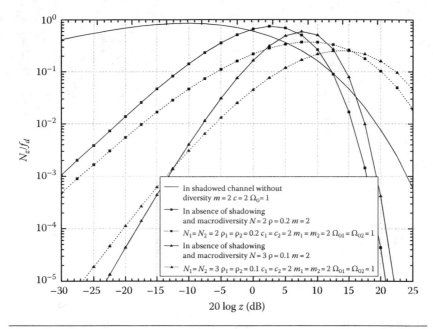

Figure 7.5 Normalized average LCR of our macrodiversity structure for various values of fading and shadowing severity level.

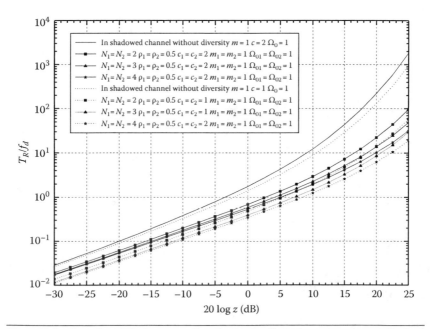

Figure 7.6 Normalized AFD of our macrodiversity structure for various values of correlation level and diversity order.

There is also a high margin of performance improvement compared to the nondiversity case. Comparison of LCR values is presented in Figure 7.6. By considering (7.6) through (7.8) and (4.13), LCR values can be obtained for the case when microdiversity is applied, considering no shadowing. Since (7.1) can be considered as the generalization of [14, Eq. (27)] for nonzero values of correlation, because it can be reduced to [14, Eq. (27)] by setting $\rho = 0$, and by adequate transformation of parameters, the obtained results are indirectly compared to those in [14]. Like in the AFD case, there is an improvement in the obtained LCR values by the proposed macrodiversity structure.

Finally, we may discuss the obtained LCR and AFD values from the point of view of [18]. By considering that the inverse Gaussian shadowed Nakagami-m composite fading channel has similar or better performances than gamma shadowed (for the value $m = 2$, AFD and LCR values for inverse Gaussian shadowing from Figures 7.1 and 7.2 are similar or better compared to gamma shadowed channel), we have shown that our model gives improvement even in non-ideal cases. In line with the previous conclusion, we may assume that this method could also be successfully applied for other composite models of fading/shadowing (not just inverse Gaussian, but also GMSM model, because by comparing with Fig. 2 from [19] a similar conclusion arises).

In summary, numerical results support the assertion that the gain in performances is made by using this macrodiversity structure. Since better results are obtained in the cases of higher values of fading/shadowing parameters m_i and c_i and lower values of correlation ρ_i, a higher number of diversity branches should be used in opposite cases, so there must be a trade-off between the desired performances and complexity.

7.1.2 Correlated Shadowing

Correlation at macrolevel is also a common phenomenon, which has been measured and shown to be significant in various wireless networks. With a single mobile station (MS) and two BSs considered at a given time, shadowing components on the two links often experience correlation, as witnessed by experimental results given in [20,21].

The level of correlation depends on the separation between the BS, on the surrounding terrain, the angle of arrival of the received signals, and various factors. It has been shown that in cellular radio systems, correlation on links between an MS and multiple BSs significantly affects mobile hand-off probabilities and cochannel interference ratios [22–24]. In addition, the coverage area and interference characteristics are affected by correlated shadowing, which occurs in digital broadcasting, links between multiple broadcast antennas to a single receiver [25]. Finally, correlated shadowing is significant (correlation coefficients even reach 0.95) and strongly impacts system performance in indoor WLANs [26]. Results obtained in the previous section will be generalized here, with correlation introduced at the macrolevel. Novel closed-form expressions will be presented for LCR and AFD. Here, long-term fading is as in [27] described with correlated gamma distributions:

$$
f_{\Omega_1\Omega_2}(\Omega_1,\Omega_2) = \frac{\rho_2^{-(1/2)(c-1)}}{\Gamma(c)(1-\rho_2)\Omega_0^{c+1}}(\Omega_1\Omega_2)^{(c-1)/2}
$$

$$
\times\exp\left(-\frac{\Omega_1+\Omega_2}{\Omega_0(1-\rho_2)}\right)I_{c-1}\left(\frac{\sqrt{4\rho_2\Omega_1\Omega_2}}{\Omega_0(1-\rho_2)}\right). \qquad (7.13)
$$

In the previous equation, shadowing correlation at the macrolevel between BSs is denoted with ρ_2. After substituting Equations 7.13, 7.7, and 7.8 into (4.13), LCR can be presented in the form of

$$
\frac{N_Z(z)}{f_d} = 2\frac{\rho_2^{-(1/2)(c-1)}}{\Gamma(c)(1-\rho_2)\Omega_0^{c+1}}\sqrt{\frac{1}{2m_1}}\frac{r^{M_1-1}}{\Gamma(M_1)}\left(\frac{N_1m_1}{r_1}\right)^{M_1}
$$

$$
\times\sum_{a=0}^{+\infty}\sum_{b=0}^{+\infty}\left\{\left(4\sqrt{\rho_2}\right)^{2b+c-1}\left(\frac{1}{\Omega_0(1-\rho_2)}\right)^{a+2b+c-1}\right.
$$

$$
\times\frac{1}{\Gamma(b+c)b!2^{2b+c-1}}(b+c)^{-1}\frac{1}{(b+c+1)_a}
$$

$$\times \left(\frac{N_1 m_1}{2r_1} \Omega_0 \left(1 - \rho_2\right) \right)^{((1/2)+a+2b+2c-M_1)/2}$$

$$\times K_{((1/2)+a+2b+2c-M_1)} \left(2\sqrt{\frac{2N_1 M_1}{r_1 \Omega_0 \left(1 - \rho_2\right)}} z \right)$$

$$+ 2 \frac{\rho_2^{-(1/2)(c-1)}}{\Gamma(c)(1-\rho_2)\Omega_0^{c+1}} \sqrt{\frac{1}{2m_2}} \frac{z^{M_2-1}}{\Gamma(M_2)} \left(\frac{N_2 m_2}{r_2} \right)^{M_2}$$

$$\times \sum_{a=0}^{+\infty} \sum_{b=0}^{+\infty} \left(4\sqrt{\rho_2}\right)^{2b+c-1} \left(\frac{1}{\Omega_0(1-\rho_2)} \right)^{a+2b+c-1}$$

$$\times \frac{1}{\Gamma(b+c)b!2^{2b+c-1}} (b+c)^{-1} \frac{1}{(b+c+1)_a}$$

$$\times \left(\frac{N_2 m_2}{2r_2} \Omega_0 (1-\rho_2) \right)^{((1/2)+a+2b+2c-M_2)/2}$$

$$\times K_{((1/2)+a+2b+2c-M_2)} \left(2\sqrt{\frac{2N_2 M_2}{r_2 \Omega_0 (1-\rho_2)}} r \right) \Bigg\}. \tag{7.14}$$

In Figure 7.7, the obtained normalized LCR values are graphically presented in the function of shadowing correlation at the macrolevel. From this figure, it is clear that lower LCR levels are crossed with the lower level of shadowing correlation between the branches, ρ_2.

7.2 SC Macrodiversity System Operating over Gamma-Shadowed κ-μ Fading Channels

In this section, closed-form expressions will be provided for the macrodiversity structure operating over the gamma-shadowed κ-μ fading channels. A macrodiversity structure of SC type consists of two microdiversity systems with switching between the BSs based on their output signal power values. Each observed microdiversity system is of MRC type with an arbitrary number of branches in the presence of

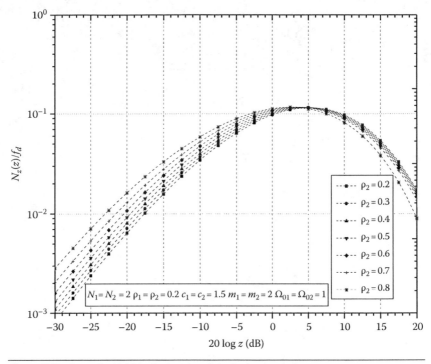

Figure 7.7 Normalized LCR in the function of shadowing correlation ρ_2.

κ-μ fading. In [28,29], it is shown that the sum of κ-μ powers is also κ-μ distributed power (but with different parameters), which is an ideal choice for MRC analysis. The expression for the PDFs of the outputs of MRC microdiversity systems is as follows [29]:

$$
f_{z_i/\Omega_i}\left(\frac{z_i}{\Omega_i}\right) = \frac{2L_i\mu_i\left(1+\kappa_i\right)^{(L_i\mu_i+1)/2}}{\kappa_i^{(L_i\mu_i-1)/2}\,e^{L_i\mu_i\kappa_i}\,\Omega_i^{(L_i\mu_i+1)/2}}\,z_i^{L_i\mu_i}\,e^{-(\mu_i(1+\kappa_i)z_i^2)/\Omega_i}
$$

$$
\times I_{L_i\mu_i-1}\left[2\mu_i\sqrt{\frac{L_i\kappa_i\left(1+\kappa_i\right)z_i^2}{\Omega_i}}\right], \tag{7.15}
$$

where

μ_i and κ_i are well-known κ-μ fading parameters of each microdiversity system

L_i denotes the number of channels at each microlevel

Considering (7.7), conditioned on Ω_i, the JPDF, $f(z_i,\dot{z}_i/\Omega_i)$, can be presented as

$$f_{z_i,\dot{z}_i/\Omega_i}\left(\frac{z_i,\dot{z}_i}{\Omega_i}\right) = \frac{2L_i\mu_i\left(1+\kappa_i\right)^{(L_i\mu_i+1)/2}}{\kappa_i^{(L_i\mu_i-1)/2}e^{L_i\mu_i\kappa_i}\Omega_i^{(L_i\mu_i+1)/2}}z_i^{L_i\mu_i}e^{-(\mu_i(1+\kappa_i)z_i^2)/\Omega_i}$$

$$\times I_{L_i\mu_i-1}\left[2\mu_i\sqrt{\frac{L_i\kappa_i\left(1+\kappa_i\right)z_i^2}{\Omega_i}}\right] \times \frac{1}{\sqrt{2\pi}\dot{\sigma}_{z_i}}\exp\left(-\frac{\dot{z}_i^2}{2\dot{\sigma}_{z_i}^2}\right). \qquad (7.16)$$

Since the outputs of MRCs and their time derivatives follow (7.3) and (7.5), then corresponding variances from (7.16) are given as [30]

$$\sigma_{\dot{z}_i}^2 = \sigma_{\dot{z}_{ik}}^2 = 2\pi^2 f_d^2 \sigma_{ik}^2, \quad \sigma_{ik}^2 = \frac{\Omega_i}{2\mu_d(1+\kappa_d)}. \qquad (7.17)$$

Considering the selection described with (7.8) and (7.9), with respect to (7.10) following the procedure explained in Appendix from [11], we can easily derive the infinite-series expression for the system output LCR in the form of

$$\frac{N_Z(z)}{f_d} = 2\sqrt{\pi}\sum_{p=0}^{\infty}\frac{L_1^p\kappa_1^p\mu_1^{2p+L_1\mu_1}(1+\kappa_1)^{p+L_1\mu_1}}{\Gamma(p+L_1\mu_1)p!\Gamma(c_1)\Gamma(c_2)\exp(L_1\mu_1\kappa_1)}z^{p+L_1\mu_1-1}$$

$$\times \sum_{q=0}^{\infty}\frac{\left((1+\kappa_1)\mu_1z/(1/\Omega_{01}+1/\Omega_{02})\right)^{(c_1+c_2+q-p-L_1\mu_1+1/2)/2}}{c_2(1+c_2)_q\Omega_{01}^{c_1}\Omega_{02}^{c_2+q}}$$

$$\times K_{(c_1+c_2+q-p-L_1\mu_1+1/2)}\left(2\sqrt{\frac{(1+\kappa_1)\mu_1z(\Omega_{01}+\Omega_{02})}{\Omega_{01}\Omega_{02}}}\right)$$

$$+2\sqrt{\pi}\sum_{p=0}^{\infty}\frac{L_2^p\kappa_2^p\mu_2^{2p+L_2\mu_2}(1+\kappa_2)^{p+L_2\mu_2}}{\Gamma(p+L_2\mu_2)p!\Gamma(c_1)\Gamma(c_2)\exp(L_2\mu_2\kappa_2)}z^{p+L_2\mu_2-1}$$

$$\times \sum_{q=0}^{\infty}\frac{\left((1+\kappa_2)\mu_2z/(1/\Omega_{01}+1/\Omega_{02})\right)^{(c_1+c_2+q-p-L_2\mu_2+1/2)/2}}{c_1(1+c_1)_q\Omega_{01}^{c_1}\Omega_{02}^{c_2+q}}$$

$$\times K_{(c_1+c_2+q-p-L_2\mu_2+1/2)}\left(2\sqrt{\frac{(1+\kappa_2)\mu_2z(\Omega_{01}+\Omega_{02})}{\Omega_{01}\Omega_{02}}}\right). \qquad (7.18)$$

Similarly, AFD can be expressed as

$$T_z(z) = \frac{F_z(z \le Z)}{N_z(z)};$$

$$F_z(z) = 2 \sum_{p=0}^{\infty} \frac{L_1^p \kappa_1^p \mu_1^p}{\Gamma(p + L_1\mu_1) \, p! \, \Gamma(c_1)\Gamma(c_2) \exp(L_1\mu_1\kappa_1)}$$

$$\times \sum_{q=0}^{\infty} \sum_{r=0}^{\infty} \left((1+\kappa_1)\mu_1 z\right)^{q+p-L_1\mu_1}$$

$$\times \frac{\left((1+\kappa_1)\mu_1 z/(1/\Omega_{01} + 1/\Omega_{02})\right)^{(c_1+c_2+r-q-p-L_1\mu_1)/2}}{c_2(1+c_2)_r \Omega_{01}{}^{c_1} \Omega_{02}{}^{c_2+r} (p+L_1\mu_1)(1+p+L_1\mu_1)_q}$$

$$\times K_{(c_1+c_2+r-q-p-L_1\mu_1)} \left(2\sqrt{\frac{(1+\kappa_1)\mu_1 z(\Omega_{01} + \Omega_{02})}{\Omega_{01}\Omega_{02}}} \right)$$

$$+ 2 \sum_{p=0}^{\infty} \frac{L_2^p \kappa_2^p \mu_2^p}{\Gamma(p + L_2\mu_2) \, p! \, \Gamma(c_1)\Gamma(c_2) \exp(L_2\mu_2\kappa_2)}$$

$$\times \sum_{q=0}^{\infty} \sum_{r=0}^{\infty} \left((1+\kappa_2)\mu_2 z\right)^{q+p-L_2\mu_2}$$

$$\times \frac{\left((1+\kappa_2)\mu_2 z/(1/\Omega_{01} + 1/\Omega_{02})\right)^{(c_1+c_2+r-q-p-L_2\mu_2)/2}}{c_1(1+c_1)_r \Omega_{01}{}^{c_1} \Omega_{02}{}^{c_2+r} (p+L_2\mu_2)(1+p+L_2\mu_2)_q}$$

$$\times K_{(c_1+c_2+r-q-p-L_2\mu_2)} \left(2\sqrt{\frac{(1+\kappa_2)\mu_2 z(\Omega_{01} + \Omega_{02})}{\Omega_{01}\Omega_{02}}} \right). \qquad (7.19)$$

The infinite series from (7.18) and (7.19) rapidly converge for any value of the parameters c_i, L_i, μ_i, and κ_i, $i = 1, 2$. In Table 7.1, the number of terms to be summed in (7.18), in order to achieve accuracy at the 5th significant digit, is presented for various values of system parameters.

Table 7.1 Terms That Need to Be Summed in Each Sum of (7.18) to Achieve Accuracy at the 5th Significant Digit

$z=-10$ dB, $c_1=c_2=1$, $\Omega_{01}=\Omega_{02}=1$		$\mu_1=\mu_2=2$	$\mu_1=\mu_2=3$
$L_1=L_2=2$	$\kappa_1=\kappa_2=0.5$	9	12
	$\kappa_1=\kappa_2=1$	12	15
$L_1=L_2=3$	$\kappa_1=\kappa_2=0.5$	12	14
	$\kappa_1=\kappa_2=1$	16	21
$z=0$ dB, $c_1=c_2=1$, $\Omega_{01}=\Omega_{02}=1$		$\mu_1=\mu_2=2$	$\mu_1=\mu_2=3$
$L_1=L_2=2$	$\kappa_1=\kappa_2=0.5$	10	12
	$\kappa_1=\kappa_2=1$	13	17
$L_1=L_2=3$	$\kappa_1=\kappa_2=0.5$	11	17
	$\kappa_1=\kappa_2=1$	17	21
$z=10$ dB, $c_1=c_2=1$, $\Omega_{01}=\Omega_{02}=1$		$\mu_1=\mu_2=2$	$\mu_1=\mu_2=3$
$L_1=L_2=2$	$\kappa_1=\kappa_2=0.5$	12	13
	$\kappa_1=\kappa_2=1$	15	18
$L_1=L_2=3$	$\kappa_1=\kappa_2=0.5$	12	15
	$\kappa_1=\kappa_2=1$	16	23

Numerically obtained results are graphically presented to examine the influence of various parameters such as shadowing and fading severity and the number of the diversity branches at the microcombiners on the concerned quantities. Normalized values of LCR, by maximal Doppler shift frequency f_d, are presented in Figures 7.8 and 7.9. We can observe from Figure 7.8 that lower levels are crossed with the higher number of diversity branches at each microcombiner and larger values of shadowing severity parameters c_i. From Figure 7.9, it is obvious that for higher values of κ-μ fading severity parameter μ_i and for higher values of dominant/scattered components power ratio κ_i, LCR values decrease, since for smaller κ and μ values, the dynamics in the channel is larger. Normalized AFD for various values of the system's parameters is presented in Figures 7.10 and 7.11. Similarly, better performances of system are achieved (lower values of AFD) with a higher number of diversity branches, higher values of fading severity, and higher values of shadowing severity.

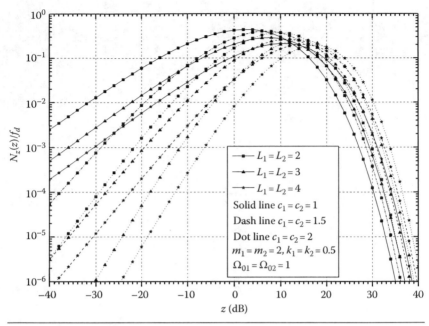

Figure 7.8 Normalized average LCR of our macrodiversity structure for various values of shadowing severity levels and diversity order.

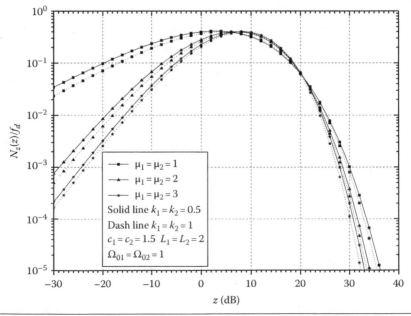

Figure 7.9 Normalized average LCR of our macrodiversity structure for various values of fading severity parameters μ and κ.

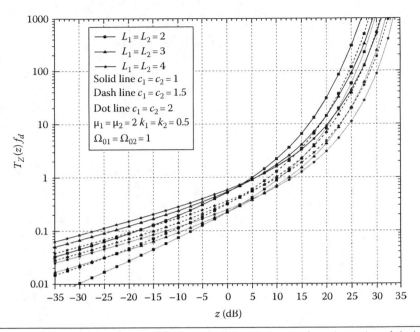

Figure 7.10 Normalized average AFD of our macrodiversity structure for various values of shadowing severity levels and diversity order.

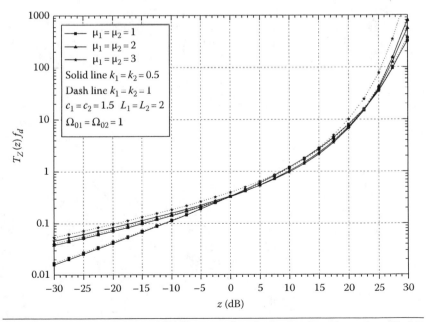

Figure 7.11 Normalized average AFD of our macrodiversity structure for various values of fading severity parameters μ and κ.

References

1. Stuber, G. L. (2003). *Mobile Communication*, 2nd edn. Kluwer Academic Publishers, Massachusetts.
2. Lee, W. C. Y. (1993). *Mobile Communications Design Fundamentals.* Wiley, New York.
3. Abu-Dayya, A. A. and Beaulieu, C. N. (1995). Micro- and macrodiversity MDPSK on shadowed frequency-selective channels. *IEEE Transactions on Communications*, 43(8), 2334–2342.
4. Adinoyi, A., Yanikomeroglu, H., and Loyka, S. (2004). Hybrid macro- and generalized selection combining microdiversity in lognormal shadowed Rayleigh fading channels. *Proceedings of IEEE International Conference on Communications*, Paris, France, Vol. 1, pp. 244–248.
5. Shankar, P. M. (2008). Analysis of microdiversity and dual channel macrodiversity in shadowed fading channels using a compound fading model. *International Journal of Electronics and Communications (AEUE)*, 62(6), 445–449.
6. Shankar, P. M. (2006). Performance analysis of diversity combining algorithms in shadowed fading channels. *Wireless Personal Communications*, 37(1), 61–72.
7. Turkmani, A. M. D. (1991). Performance evaluation of composite microscopic plus macroscopic diversity system. *IEE Proceedings, Part I: Communications, Speech and Vision*, 138, 15–20.
8. Jeong W. C. and Chung J. M. (2005). Analysis of macroscopic diversity combining of MIMO signals in mobile communications. *International Journal of Electronics and Communications (AEUE)*, 59, 454–462.
9. Al-Hussaini, E. K., et al. (2002). Composite macroscopic and microscopic diversity of sectorized macrocellular and microcellular mobile radio systems employing RAKE receiver over Nakagami fading plus lognormal shadowing channel. *Wireless Personal Communications*, 21(3), 309–328.
10. Stefanovic, D., Panic, S., and Spalevic, P. (2011). Second order statistics of SC macrodiversity system operating over gamma shadowed Nakagami-m fading channels. *International Journal of Electronics and Communications (AEUE)*, 65(5), 413–418.
11. Panic, S. et al. (2011). Second-order statistics of selection macrodiversity system operating over gamma shadowed κ-μ fading channels. *EURASIP Journal on Wireless Communications and Networking*, 2011, Article ID 151, 1–7.
12. Mukherjee, S. and Avidor, D. (2003). Effect of microdiversity and correlated macrodiversity on outages in a cellular system. *IEEE Transactions on Wireless Technology*, 2(1), 50–59.
13. Aalo, V. A. (1995). Performance of maximal-ratio diversity systems in a correlated Nakagami fading environment. *IEEE Transactions on Communications*, 43(8), 2360–2369.

14. Iskander, C. D. and Mathiopoulos, P. T. (2002). Analytical level crossing rate and average fade duration in Nakagami fading channels. *IEEE Transactions on Communications*, 50(8), 1301–1309.
15. Dong, X. and Beaulieu, N. C. (2001). Average level crossing rate and average fade duration of selection diversity. *IEEE Communication Letters*, 10(5), 396–399.
16. Shankar, P. M. (2007). Outage analysis in wireless channels with multiple interferers subject to shadowing and fading using a compound pdf model. *International Journal of Electronics and Communications (AEUE)*, 61(4), 255–261.
17. Vetelino, F. S., Young, C., and Andrews, L. (2007). Fade statistics and aperture averaging for Gaussian beam waves in moderate-to-strong turbulence. *Applied Optics*, 46(18), 3780–3789.
18. Trigui, I. et al. (2008). On the level crossing rate and average fade duration of composite multipath/shadowing channels. *Proceedings of IEEE Global Telecommunications Conference*, New Orleans, LA, pp. 1–5.
19. Ramos, F. A., Kontorovitch, V., and Lara, M. (2005). On the generalized and modified Suzuki model (GMSM): Approximations and level crossing statistics. *Proceedings of International Conference on Electrical and Electronics Engineering*, Mexico City, Mexico, pp. 110–113.
20. Graziano, V. (1978). Propagation correlations at 900 MHz. *IEEE Transactions on Vehicular Technology*, 27(4), 182–189.
21. Van Rees, J. (1987). Cochannel measurements for interference limited small cell planning. *Archiv der elektrischen Übertragung*, 41, 318–320.
22. Klingenbrunn, T. and Mogensen, P. (1999). Modelling cross-correlated shadowing in network simulations. *Proceedings of IEEE Vehicular Technology Conference*, Amsterdam, the Netherlands, Vol. 3, pp. 1407–1411.
23. Zhang, J. and Aalo, V. (2001). Effect of macrodiversity on average-error probabilities in a Rician fading channel with correlated lognormal shadowing. *IEEE Transactions on Communications*, 49(1), 14–18.
24. Safak, A. and Prasad, R. (1991). Effects of correlated shadowing signals on channel reuse in mobile radio systems. *IEEE Transactions on Vehicular Technology*, 40(4), 708–713.
25. Malmgren, G. (1997). On the performance of single frequency networks in correlated shadow fading. *IEEE Transactions on Broadcasting*, 43(2), 155–165.
26. Butterworth, K. S., Sowerby, K. W., and Williamson, A. G. (2000). Base station placement for in-building mobile communication systems to yield high capacity and efficiency. *IEEE Transactions on Communications*, 48(4), 658–669.
27. Sekulovic, N. and Stefanovic, M. (2012). Performance analysis of system with micro- and macrodiversity reception in correlated gamma shadowed Rician fading channels. *Wireless Personal Communications*, 65(1), 143–156.
28. Yacoub, M. D. (2007). The η-μ distribution and the κ-μ distribution. *IEEE Antennas and Propagation Magazine*, 49(1), 68–81.

29. Milisic, M., Hamza, M., and Hadzialic M. (2009). BEP/SEP and outage performance analysis of L-branch maximal-ratio combiner for κ-μ fading. *International Journal of Digital Multimedia Broadcasting*, 2009, Article ID 573404, 1–8.
30. Cotton, S. L. and Scanlon, W. G. (2007). Higher-order statistics for kappa-mu distribution. *Electronic Letters*, 43(22), 1215–1217.

8

EVALUATIONS OF CHANNEL CAPACITY UNDER VARIOUS ADAPTATION POLICIES AND DIVERSITY TECHNIQUES

Recently, the expansion of wireless services has required more spectrally efficient communication to meet the consumer demand. So the primary concern in future wireless communications systems is for conserving, sharing, and using bandwidth efficiently. Therefore, channel capacity is one of the most important concerns in the design of wireless systems, as it determines the maximum attainable throughput of the system [1]. It can be defined as the average transmitted data rate per unit bandwidth, for a specified average transmit power, and specified level of received outage or bit-error rate [2]. A Skilful combination of bandwidth-efficient coding and modulation schemes can be used to achieve higher channel capacities per unit bandwidth. However, mobile radio links are, due to the combination of randomly delayed reflected, scattered, and diffracted signal components, subjected to severe multipath fading, which leads to serious degradation in the link signal-to-noise ratio (SNR). An effective scheme that can be used to overcome fading influence is adaptive transmission. The performance of adaptation schemes is further improved by combining them with space diversity, since diversity combining is a powerful technique that can be used to combat fading in wireless systems resulting in improved link performance [3].

8.1 Channel and System Model

Diversity combining is a powerful technique that can be used to combat fading effect in wireless systems [4]. The optimal diversity combining technique is the maximum ratio combining (MRC) technique.

This combining technique involves cophasing of the useful signal in all branches, multiplication of the received signal in each branch by a weight factor that is proportional to the estimated ratio of the envelope and the power of that particular signal, and summing of the received signals from all antennas. By cophasing, all the random phase fluctuations of the signal, which emerged during the transmission, are eliminated. For this process, it is necessary to estimate the phase of the received signal, so this technique requires the entire amount of the channel state information (CSI) of the received signal and a separate receiver chain for each branch of the diversity system, which increases the complexity of the system [5].

One of the least complicated combining methods is selection combining (SC). While other combining techniques require all or some of the amount of the CSI of the received signal and a separate receiver chain for each branch of the diversity system, which increase its complexity, the SC receiver processes only one of the diversity branches and is much simpler for practical realization, in contrast to these combining techniques [4–7]. Generally, SC selects the branch with the highest SNR, that is, the branch with the strongest signal, assuming that noise power is equally distributed over branches. Since receiver diversity mitigates the impact of fading, the question is whether it also increases the capacity of a fading channel.

Another effective scheme that can be used to overcome fading influence is adaptive transmission. Adaptive transmission is based on the receiver's estimation of the channel and feedback of the CSI to the transmitter. The transmitter then adapts the transmit power level, symbol/bit rate, constellation size, coding rate/scheme, or any combination of these parameters in response to the changing of channel conditions [8]. Adapting certain parameters of the transmitted signal to the fading channel can help better utilization of the channel capacity. These transmissions provide much higher channel capacities per unit bandwidth by taking advantage of favorable propagation conditions: transmitting at high speeds under favorable channel conditions and responding to channel degradation through a smooth reduction of their data throughput. The source may transmit faster and/or at a higher power under good channel conditions and slower and/or at a reduced power under poor conditions. A reliable feedback path between that estimator and the transmitter and accurate channel estimation at the receiver is required for achieving good performances of adaptive transmission. Widely accepted adaptation policies include optimal power and rate adaptation (OPRA), constant power with optimal rate adaptation (ORA), channel inversion with fixed rate (CIFR), and truncated CIFR (TIFR). Results obtained for this protocols show the trade-off between capacity and complexity. The adaptive policy with transmitter and receiver side information requires more complexity in the transmitter (and it typically also requires a feedback path between the receiver and transmitter to obtain the side information). However, the decoder in the receiver is relatively simple. The nonadaptive policy has a relatively simple transmission scheme, but its code design must use the channel correlation statistics (often unknown), and the decoder complexity is proportional to the channel decorrelation time. The channel inversion and truncated inversion policies use codes designed for the additive white Gaussian noise (AWGN) channels and are therefore the least complex to implement, but in severe fading conditions they exhibit large capacity losses relative to the other techniques.

The performance of adaptation schemes is further improved by combining them with space diversity. The hypothesis that the variation of the combined output SNR is tracked perfectly by the receiver and that the variation in SNR is sent back to the transmitter via an error-free feedback path will be assumed in the ongoing analysis [8]. In addition, it is assumed that the time delay in this feedback path is negligible compared to the rate of the channel variation. Following these assumptions, the transmitter could adapt its power and/or rate relative to the actual channel state.

There are numerous published articles based on the study of channel capacity evaluation. In [9], the capacity of Rayleigh fading channels under four adaptation policies and a multibranch system with variable correlation is investigated. The capacity of Rayleigh fading channels under different adaptive transmission and different diversity combining techniques is also studied in [7,10]. In [11], the channel capacity of MRC over exponentially correlated Nakagami-m fading channels under adaptive transmission is analyzed. The channel capacity of adaptive transmission schemes using an equal gain combining (EGC) receiver over Hoyt fading channels is presented in [12].

In [13], dual-branch SC receivers operating over correlative Weibull fading under three adaptation policies are analyzed.

In this chapter, we will focus on more general and nonlinear fading distributions. We will perform an analytical study of the κ-μ fading channel capacity under the OPRA, ORA, CIFR, and TIFR adaptation policies and MRC and SC diversity techniques. To the best of the authors' knowledge, such a study has not been previously considered in the open technical literature. The expressions for the proposed adaptation policies and diversity techniques will be derived. Capitalizing on them, numerically obtained results will be graphically presented, in order to show the effects of various system parameters, such as diversity order and fading severity on observed performances. In a similar manner, an analytical study of the Weibull fading channel capacity, under the OPRA, ORA, CIFR, and TIFR adaptation policies and MRC diversity technique will be performed.

8.1.1 κ-μ *Fading Channel and System Model*

The multipath fading in wireless communications is modeled by several distributions including Nakagami-*m*, Hoyt, Rayleigh, and Rice. By considering important phenomena inherent in radio propagation, the κ-μ fading model was recently proposed in [14] as a fading model, which describes the short-term signal variation in the presence of line-of-sight (LOS) components. This distribution is more realistic than other special distributions, since its derivation is completely based on a nonhomogeneous scattering environment. Also κ-μ as a general physical fading model includes Rayleigh, Rician, and Nakagami-*m* fading models, as its special cases [14]. The κ-μ distribution is written in terms of two physical parameters, κ and μ. The parameter κ is related to the multipath clustering and the parameter μ is the ratio between the total power of the dominant components and the total power of the scattered waves. In the case of κ = 0, the κ-μ distribution is equivalent to the Nakagami-*m* distribution. When μ = 1, the κ-μ distribution becomes the Rician distribution with κ as the Rice factor. Moreover, the κ-μ distribution fully describes the characteristics of the fading signal in terms of measurable physical parameters.

The SNR in the κ-μ fading channel follows the probability density function (PDF) given by [15]

$$f_\gamma(\gamma) = \frac{\mu}{k^{(\mu-1)/2}e^{\mu k}}\left(\frac{1+k}{\overline{\gamma}}\right)^{(\mu+1)/2}\gamma^{(\mu-1)/2}\,e^{-\mu(1+k)\gamma/\overline{\gamma}}I_{\mu-1}\left(2\mu\sqrt{\frac{(1+k)k\gamma}{\overline{\gamma}}}\right).$$

(8.1)

In the previous equation

$\overline{\gamma}$ is the corresponding average SNR

$I_n(x)$ denotes the nth-order-modified Bessel function of the first kind [16]

κ and μ are well-known κ-μ fading parameters

Using the series representation of Bessel function [16, Eq. (8.445)]

$$I_n(x) = \sum_{k=0}^{+\infty}\frac{x^{2k+n}}{2^{2k+n}\Gamma(k+n+1)k!},$$

(8.2)

the cumulative distribution function (CDF) of γ can be written in the form of

$$F_\gamma(\gamma) = \sum_{p=0}^{+\infty}\frac{\mu^p\kappa^p}{e^{\mu\kappa}\Gamma(p+\mu)}\Lambda\left(p+\mu,\frac{\mu(1+\kappa)\gamma}{\overline{\gamma}}\right),$$

(8.3)

with $\Gamma(x)$ and $\Lambda(a,x)$ denoting gamma and lower incomplete gamma function, respectively [16, Eqs. (8.310.1, 8.350.1)].

It is shown in [15], that the sum of κ-μ squares is κ-μ square as well but with different parameters, which is an ideal choice for MRC analysis. Then the expression for the PDF of the outputs of MRC diversity systems is as follows [15, Eq. (11)]:

$$f_\gamma^{MRC}(\gamma) = \frac{L\mu}{k^{(L\mu-1)/2}e^{L\mu k}}\left(\frac{1+k}{L\overline{\gamma}}\right)^{(L\mu+1)/2}$$

$$\times\gamma^{(L\mu-1)/2}e^{-\mu(1+k)\gamma/\Omega}I_{L\mu-1}\left(2\mu L\sqrt{\frac{(1+k)k\gamma}{L\overline{\gamma}}}\right),$$

(8.4)

with L denoting the number of diversity branches.

Furthermore, the expression for the PDF of outputs of SC diversity system can be obtained by substituting expressions (8.1) and (8.3) into

$$f_\gamma^{SC}(\gamma) = \sum_{i=1}^{L} f_{\gamma_i}(\gamma) \prod_{\substack{j=1 \\ j \neq i}}^{L} F_{\gamma_j}(\gamma), \qquad (8.5)$$

where

$f_{\gamma_i}(\gamma)$ and $F_{\gamma_i}(\gamma)$ define the PDF and CDF of SNRs at input branches, respectively

L denotes the number of diversity branches

8.1.2 Weibull Fading Channel and System Model

The previously mentioned well-known fading distributions are derived assuming a homogeneous diffuse scattering field, resulting from randomly distributed point scatterers. The assumption of a homogeneous diffuse scattering field is certainly an approximation, because the surfaces are spatially correlated characterizing a nonlinear environment. With the aim to explore the nonlinearity of the propagation medium, a general fading distribution, the Weibull distribution, was proposed. The nonlinearity is manifested in terms of a power parameter $\beta > 0$, such that the resulting signal intensity is obtained not simply as the modulus of the multipath component but as the modulus to a certain given power. As β increases, the fading severity decreases, while for the special case of $\beta = 2$ reduces to the well-known Rayleigh distribution. Weibull distribution seems to exhibit good fit to experimental fading channel measurements, for both indoor and outdoor environments.

The SNR in a Weibull fading channel follows the PDF given by [17, Eq. (14)]:

$$f(\gamma) = \frac{\beta}{2a\bar{\gamma}} \left(\frac{\gamma}{a\bar{\gamma}} \right)^{(\beta/2)-1} e^{-(\gamma/a\bar{\gamma})^{\beta/2}}, \qquad (8.6)$$

where

$\bar{\gamma}$ is the corresponding average SNR

β is well-known Weibull fading parameter

$a = 1/\Gamma(1 + 2/\beta)$

It is shown in [18,19] that the expression for the PDF of the outputs of MRC diversity systems follows [19, Eq. (1)]:

$$f_\gamma^{MRC}(\gamma) = \frac{\beta\gamma^{L\beta/2-1}}{2\Gamma(L)(\Xi\bar\gamma)^{L\beta/2}} e^{-(\gamma/\Xi\bar\gamma)^{\beta/2}}; \quad \Xi = \frac{\Gamma(L)}{\Gamma(L+2/\beta)}, \quad (8.7)$$

with L denoting the number of diversity branches. Similarly, expression for the PDF of the outputs of SC diversity systems can be obtained using (8.5).

8.2 Optimal Power and Rate Adaptation Policy

In the OPRA protocol, the power level and rate parameters vary in response to the changing channel conditions. It achieves the ergodic capacity of the system, that is, the maximum achievable average rate by use of adaptive transmission. However, OPRA is not suitable for all applications because for some of them it requires a fixed rate.

During our analysis, it is assumed that the variation in the combined output SNR, γ, over κ-μ fading channels is tracked perfectly by the receiver and that variation of γ is sent back to the transmitter via an error-fee feedback path. Compared to the rate of channel variation, the time delay in this feedback is negligible. These assumptions allow the transmitter to adopt its power and rate correspondingly to the actual channel state. Channel capacity of the fading channel with received SNR distribution, $p_\gamma(\gamma)$, under optimal power and rate adaptation policy, is given by [8]

$$\langle C \rangle_{opra} = B \int_{\gamma_0}^{\infty} \log_2\left(\frac{\gamma}{\gamma_0}\right) f_\gamma(\gamma) d\gamma, \quad (8.8)$$

where
  B (Hz) denotes the channel bandwidth
  γ_0 is the SNR cutoff level below which transmission of data is suspended

This cutoff level must satisfy the following equation:

$$\int_{\gamma_0}^{\infty}\left(\frac{1}{\gamma_0}-\frac{1}{\gamma}\right)f_\gamma(\gamma)d\gamma = 1. \tag{8.9}$$

Since no data are sent when $\gamma < \gamma_0$, the optimal policy suffers a probability of outage equal to the probability of no transmission given by

$$P_{out} = \int_0^{\gamma_0} f_\gamma(\gamma)d\gamma = 1 - \int_{\gamma_0}^{\infty} f_\gamma(\gamma)d\gamma. \tag{8.10}$$

8.2.1 κ-μ *Fading Channels*

To achieve the capacity in (8.8), the channel fading level must be attended at the receiver as well as at the transmitter. The transmitter has to adapt its power and rate to the actual channel state: when γ is large, high power levels and rates are allocated for good channel conditions, and lower power levels and rates for unfavorable channel conditions when γ is small. Substituting (8.1) into (8.9), we found that the cutoff level must satisfy

$$\sum_{i=0}^{\infty}\frac{(kL\mu)^i}{e^{L\mu k}\Gamma(i+L\mu)i!}\left(\frac{1}{\gamma_0}\Lambda\left(L\mu+i,\frac{\mu(1+k)\gamma_0}{\bar{\gamma}}\right)\right.$$
$$\left.-\frac{\mu(1+k)}{\bar{\gamma}}\Lambda\left(L\mu+i-1,\frac{\mu(1+k)\gamma_0}{\bar{\gamma}}\right)\right)-1=0. \tag{8.11}$$

Substituting (8.4) into (8.8), we obtain the capacity per unit bandwidth, $\langle C\rangle_{opra}/B$, when MRC diversity reception is applied, as

$$\frac{\langle C\rangle_{opra}^{MRC}}{B} = \sum_{i=0}^{\infty}\frac{L\mu}{k^{(L\mu-1)/2}e^{L\mu k}}\left(\frac{1+k}{L\bar{\gamma}}\right)^{(L\mu+1)/2}$$
$$\times \int_{\gamma_0}^{\infty}\log_2\left(\frac{\gamma}{\gamma_0}\right)\gamma^{L\mu+i-1}e^{-\mu(1+k)\gamma/\bar{\gamma}}d\gamma. \tag{8.12}$$

Now, by making change of variables and after some manipulations, the previous expression can be rewritten as

$$\frac{\langle C\rangle_{opra}^{MRC}}{B} = \sum_{i=0}^{\infty} \frac{(L\mu k)^i}{\Gamma(i+L\mu)i!e^{L\mu k}} \left(\int_0^{\infty} \log_2\left(\frac{t\bar{\gamma}}{\mu(1+k)\gamma_0}\right) t^{L\mu+i-1} e^{-t} dt \right.$$

$$\left. - \int_0^{\gamma_0\mu(1+k)/\bar{\gamma}} \log_2\left(\frac{t\bar{\gamma}}{\mu(1+k)\gamma_0}\right) t^{L\mu+i-1} e^{-t} dt \right)$$

$$= \sum_{i=0}^{\infty} \frac{(L\mu k)^i}{\Gamma(i+L\mu)i!e^{L\mu k}} (I_1 - I_2). \tag{8.13}$$

Integral I_1 can be solved by applying Gauss–Laguerre quadrature formulae in the following way:

$$I_1 = \int_0^{\infty} f_1(t)e^{-t} dt \cong \sum_{k=1}^{R} A_k f_1(t_k); \quad f_1(t) = \log_2\left(\frac{t\bar{\gamma}}{\mu(1+k)\gamma_0}\right) t^{L\mu+i-1}, \tag{8.14}$$

where A_k and t_k, $k=1,2, \ldots, R$, are, respectively, weights and nodes of Laguerre polynomials [20, pp. 875–924]. Similarly, integral I_2 can be solved by applying Gauss–Legendre quadrature formulae:

$$I_2 = \left(\frac{\gamma_0\mu(1+k)}{2\bar{\gamma}}\right)^{L\mu+i} \int_{-1}^{1} f_2(u) du \cong \left(\frac{\gamma_0\mu(1+k)}{2\bar{\gamma}}\right)^{L\mu+i} \sum_{k=1}^{R} B_k f_2(u_k), \tag{8.15}$$

where B_k and u_k, $k=1,2, \ldots, R$, are respectively weights and nodes of Legendre polynomials.

The convergence of infinite series expressions in (8.13) is rapid since we need about 10 terms to be summed in order to achieve accuracy at the 5th significant digit for corresponding values of system parameters.

8.2.2 Weibull Fading Channels

Substituting (8.7) in (8.8), the integral of the following form needs to be solved:

$$I = \frac{1}{\ln 2} \int_{\gamma_0}^{\infty} \gamma^{L\beta/2-1} \ln\left(\frac{\gamma}{\gamma_0}\right) e^{-(\gamma/\Xi\bar{\gamma})^{\beta/2}} d\gamma. \tag{8.16}$$

After making a change in variables, $t = (\gamma/\gamma_0)^{\beta/2}$, and some simple mathematical manipulations, we get

$$I = \frac{4\gamma_0^{L\beta/2}}{\beta^2 \ln 2} \int_{1}^{\infty} t^{L-1} \ln(t) e^{-(\gamma_0/\Xi\bar{\gamma})^{\beta/2}t} dt. \tag{8.17}$$

Furthermore, this integral can be evaluated using partial integration:

$$\int_{1}^{\infty} u \, d\upsilon = \lim_{t \to \infty}(u\upsilon) - \lim_{t \to 1}(u\upsilon) - \int_{1}^{+\infty} \upsilon \, du, \tag{8.18}$$

with respect to

$$u = \ln t; \quad d\upsilon = t^{L-1} e^{-(\gamma_0/\Xi\bar{\gamma})^{\beta/2}t} dt. \tag{8.19}$$

Furthermore, performing $L - 1$ successive integration by parts [16, Eq. (2.321.2)], we get

$$\upsilon = -e^{-mt} \sum_{p=1}^{L} \frac{(L-1)!}{(L-p)!} \frac{t^{L-p}}{m^p}; \quad m = \left(\frac{\gamma_0}{\Xi\bar{\gamma}}\right)^{\beta/2}. \tag{8.20}$$

Substituting (8.20) in (8.18), we see that the first two terms tend to zero. Hence, the integral in (8.17) can be solved in closed form using [16, Eq. (3.381.3)]

$$I = \frac{(L-1)!}{m^L} \sum_{p=0}^{L-1} \frac{\Gamma(p,m)}{p!}, \tag{8.21}$$

with $\Gamma(a, x)$ as higher incomplete gamma function [16]. Finally, $\langle C \rangle_{opra}/B$ using L-branch MRC diversity receiver over Weibull fading channels has this form:

$$\frac{\langle C \rangle_{opra}^{MRC}}{B} = \frac{2}{\beta \ln 2} \sum_{p=0}^{L-1} \frac{\Gamma(p,m)}{p!}. \tag{8.22}$$

8.3 Constant Power with Optimal Rate Adaptation Policy

Using ORA protocol, the transmitter adapts its rate only while maintaining a fixed power level. Thus, this protocol can be implemented at reduced complexity and is more practical than that of optimal simultaneous power and rate adaptation.

The channel capacity, $\langle C \rangle_{ora}$, with constant transmit power policy is given by [1]

$$\langle C \rangle_{ora} = B \int_0^{\infty} \log_2 (1+\gamma) f_\gamma(\gamma) d\gamma. \tag{8.23}$$

8.3.1 κ-μ Fading Channels

To achieve the capacity in (8.23), the channel fading level must be attended at the receiver as well as at the transmitter.

After substituting (8.1) into (8.23), by using partial integration

$$\int_0^{\infty} u \, dv = \lim_{\gamma \to \infty}(uv) - \lim_{\gamma \to 0}(uv) - \int_0^{\infty} v \, du;$$

$$u = \ln(1+\gamma); \quad du = \frac{d\gamma}{1+\gamma}; \quad dv = \gamma^{p+\mu-1} e^{-\mu(1+k)\gamma/\bar{\gamma}}, \tag{8.24}$$

and performing successive integration by parts [16, Eq. (2.321.2)], we get

$$v = e^{-\mu(1+k)\gamma/\bar{\gamma}} \sum_{q=1}^{p+\mu} \frac{(p+\mu-1)! \gamma^{p+\mu-k}}{(p+\mu-q)!} \left(\frac{\bar{\gamma}}{\mu(1+k)}\right)^q. \tag{8.25}$$

By substituting (8.25) in (8.24), we see that the first two terms tend to zero. Hence, the integral in (8.24) can be solved in closed form using [16, Eq. (3.381.3)]. Finally, $\langle C \rangle_{ora}/B$ over κ-μ fading channels has this form:

$$
\langle C \rangle_{ora} = \frac{B}{\ln 2} \sum_{p=0}^{\infty} \sum_{q=1}^{p+\mu} \frac{\mu^{2p+\mu-q} \kappa^p (1+\kappa)^{p+\mu-q} (n-1)!}{e^{\mu\kappa} \overline{\gamma}^{p+\mu-q} \Gamma(p+\mu) p!} e^{\mu(1+\kappa)/\overline{\gamma}}
$$

$$
\times \Gamma\left(-n+p+\mu, \frac{\mu(1+\kappa)}{\overline{\gamma}}\right). \tag{8.26}
$$

On the other hand, substituting (8.4) into (8.22) and applying a similar procedure, the expression for the $\langle C \rangle_{ora}/B$ with MRC diversity reception is derived as

$$
\langle C \rangle_{ora}^{MRC} = \frac{B}{\ln 2} \sum_{p=0}^{\infty} \sum_{q=1}^{p+\mu} \frac{\mu^{2p+\mu-q} \kappa^p (1+\kappa)^{p+\mu-q} (n-1)! L^p}{e^{L\mu\kappa} \overline{\gamma}^{p+\mu-q} \Gamma(p+\mu) p!} e^{\mu(1+\kappa)/\overline{\gamma}}
$$

$$
\times \Gamma\left(-n+p+\mu L, \frac{\mu(1+\kappa)}{\overline{\gamma}}\right). \tag{8.27}
$$

Convergence of infinite series expressions in (8.26) and (8.27) is rapid, since we need 5–10 terms to be summed in order to achieve accuracy at the 5th significant digit for corresponding values of system parameters.

8.3.2 Weibull Fading Channels

After substituting (8.6) into (8.22), when MRC reception is applied over the Weibull fading channel, we can obtain the expression for the ORA channel capacity, in the following form:

$$
\frac{\langle C \rangle_{ora}}{B} = \frac{\beta}{2\Gamma(L)(\Xi\overline{\gamma})^{L\beta/2} \ln 2} \int_0^{\infty} \gamma^{L\beta/2} \ln(1+\gamma) e^{-(\gamma/\Xi\overline{\gamma})^{\beta/2}} d\gamma. \tag{8.28}
$$

By expressing the logarithmic and exponential integrands as Meijer's G-functions [21, Eq. (11)] and using [22, Eq. (07.34.21.0012.01)], the integral in (8.28) is solved in closed form as

$$\frac{\langle C \rangle_{ora}}{B} = \frac{\beta}{2\Gamma(L)(\Xi\bar{\gamma})^{L\beta/2}\ln 2}$$

$$\times H_{2,3}^{3,1}\left((\Xi\bar{\gamma})^{-\beta/2}\left|\begin{array}{c}(-L\beta/2,\beta/2),(1-L\beta/2,\beta/2)\\(0,1),(-L\beta/2,\beta/2),(-L\beta/2,\beta/2)\end{array}\right.\right), \quad (8.29)$$

with

$$H_{p,q}^{m,n}\left(x\left|\begin{array}{c}(a_1,\alpha_1)\cdots(a_p,\alpha_p)\\(b_1,\beta_q)\cdots(b_p,\beta_q)\end{array}\right.\right) \qquad (8.30)$$

denoting the Fox's H function [23].

8.4 Channel Inversion with Fixed Rate Adaptation Policy

Channel inversion with fixed rate policy is quite different from the first two protocols as it maintains a constant rate and adapts its power to the inverse of the channels fading. The CIFR protocol achieves what is known as the outage capacity of the system. That is the maximum constant data rate that can be supported for all channel conditions with some probability of outage. However, the capacity of channel inversion is always less than the capacity of the previous two protocols as the transmission rate is fixed. On the other hand, constant rate transmission is required in some applications and is worth the loss in achievable capacity. The CIFR is an adaptation technique based on inverting the channel fading. It is the least complex technique to implement assuming that the transmitter in this way adapts its power to maintain a constant SNR at the receiver. Since a large amount of the transmitted power is required to compensate for the deep channel fades, channel inversion with fixed rate suffers a certain capacity penalty compared to the other techniques.

The channel capacity with this technique is derived from the capacity of an AWGN channel and is given as [8]

$$\langle C \rangle_{cifr} = B\log_2\left(1 + \frac{1}{\int_0^\infty (f_\gamma(\gamma)/\gamma)d\gamma}\right). \qquad (8.31)$$

8.4.1 κ-μ Fading Channels

After substituting (8.1) into (8.31), and by using [16, Eq. (6.643.2)], we get

$$\int_0^\infty x^{\mu-(1/2)} e^{-\alpha x} I_{2v}\left(2\beta\sqrt{x}\right) dx$$

$$= \frac{\Gamma\left(\mu+v+(1/2)\right)}{\Gamma(2v+1)} \beta^{-1} e^{-\beta^2/2\alpha} \alpha^{-v} M_{-\mu,v}\left(\frac{\beta^2}{\alpha}\right), \qquad (8.32)$$

where $M_{k,m}(x)$ is the Wittaker's function. We can obtain the expression for the CIFR channel capacity in the following form:

$$\langle C \rangle_{cifr} = B\log_2\left(1+\frac{(\mu-1)}{e^{-\mu k/2}\left((1+k)/k\bar\gamma\right)^{\mu/2} M_{1-(\mu/2),(\mu-1)/2}\left(\mu k\right)}\right). \qquad (8.33)$$

The case when MRC diversity is applied can be modeled by

$$\langle C \rangle_{cifr}^{MRC} = B\log_2\left(1+\frac{(L\mu-1)}{e^{-\mu kL/2}\left((1+k)/kL\bar\gamma\right)^{L\mu/2} M_{1-(L\mu/2),(L\mu-1)/2}\left(\mu kL\right)}\right). \qquad (8.34)$$

Similarly, after substituting (8.5) into (8.31), with respect to [16, Eqs. (8.531, 7.552.5, 9.14)], respectively,

$$\Lambda(a,x) = \frac{x^a}{a} e^{-x} \,_1F_1(1;1+a;x), \qquad (8.35)$$

$$\int_0^\infty e^{-x} x^{s-1} \,_pF_q(a_1,\ldots,a_p;b_1,\ldots,b_q;\alpha x) dx$$

$$= \Gamma(s) \,_{p+1}F_q(s,a_1,\ldots,a_p;b_1,\ldots,b_q;\alpha x), \qquad (8.36)$$

$$_1F_1(a;b;x) = \sum_{k=0}^\infty \frac{(a)_k x^k}{(b)_k k!} \qquad (8.37)$$

expressions for the CIFR channel capacity over κ-μ fading with SC diversity reception for dual- and triple-branch combining can be obtained in the following forms:

$$\langle C \rangle_{cifr}^{SC-2} = B\log_2\left(1 + 1\bigg/ \sum_{p=0}^{\infty}\sum_{q=0}^{\infty} f_1\right)$$

$$f_1 = \frac{\mu^{p+q+1}\kappa^{p+q}(1+\kappa)\Gamma(p+q+2\mu-1)}{2^{p+q+2\mu-2}e^{2\mu\kappa}\overline{\gamma}\Gamma(p+\mu)p!\Gamma(q+\mu)q!(q+\mu)}$$

$$\times {}_2F_1\left(p+q+2\mu-1,1;1+q+\mu;\frac{1}{2}\right) \tag{8.38}$$

$$\langle C \rangle_{cifr}^{SC-3} = B\log_2\left(1 + 1\bigg/ \sum_{p=0}^{\infty}\sum_{q=0}^{\infty}\sum_{r=0}^{\infty}\sum_{s=0}^{\infty} f_2\right)$$

$$f_2 = \frac{\mu^{p+q+r+1}\kappa^{p+q+r}(1+\kappa)\Gamma(p+q+r+s+3\mu-1)}{3^{p+q+r+s+3\mu-3}e^{3\mu\kappa}\overline{\gamma}\Gamma(p+\mu)p!\Gamma(q+\mu)q!(q+\mu)\Gamma(r+\mu)r!(r+\mu)(1+r+\mu),}$$

$$\times {}_2F_1\left(p+q+r+s+3\mu-1,1;1+q+\mu;\frac{1}{3}\right). \tag{8.39}$$

The number of terms that need to be summed in (8.38) and (8.39) to achieve accuracy at the 5th significant digit for some values of system parameters is presented in Table 8.1.

8.4.2 Weibull Fading Channels

After substituting (8.6) into (8.31), we can obtain the expression for the CIFR channel capacity when MRC diversity is applied in the following form:

Table 8.1 Number of Terms That Need to Be Summed in (8.38) and (8.39) to Achieve Accuracy at the Specified Significant Digit for Some Values of System Parameters

		$\overline{\gamma} = 5$ dB	$\overline{\gamma} = 10$ dB	$\overline{\gamma} = 15$ dB
EXPRESSION (8.38) 6TH SIGNIFICANT DIGIT				
$\kappa=1$	$\mu=1$	8	9	10
$\kappa=2$	$\mu=2$	15	15	16
EXPRESSION (8.39) 6TH SIGNIFICANT DIGIT				
$\kappa=1$	$\mu=1$	19	21	24
$\kappa=2$	$\mu=2$	23	26	28

$$\frac{\langle C \rangle_{cifr}^{MRC}}{B} = \log_2\left(1 + \frac{(\Xi\bar{\gamma})\Gamma(L)}{\Gamma(L - 2/\beta)}\right). \tag{8.40}$$

8.5 Truncated Channel Inversion with Fixed Rate

The channel inversion and truncated inversion policies use codes designed for AWGN channels and are therefore the least complex to implement, but in severe fading conditions they exhibit large capacity losses relative to the other techniques.

The truncated channel inversion policy inverts the channel fading only above a fixed cutoff fade depth γ_0. The capacity with this truncated channel inversion and fixed rate policy, $\langle C \rangle_{tifr}/B$, can be evaluated as [8]

$$\langle C \rangle_{tifr} = B \log_2\left(1 + \frac{1}{\int_{\gamma_0}^{\infty} (f_\gamma(\gamma)/\gamma)\,d\gamma}\right)(1 - P_{out}). \tag{8.41}$$

8.5.1 κ-μ Fading Channels

After substituting (8.1) into (8.41), with respect to (8.2), we can obtain the expression for evaluating CIFR channel capacity over the κ-μ fading channel in the following form:

$$\langle C \rangle_{tifr} = B \log_2\left(1 + \frac{1}{\sum_{p=0}^{\infty} f_3}\right)$$

$$\times \left(1 - \sum_{i=0}^{\infty} \frac{(k\mu)^i}{e^{\mu k}\Gamma(i+\mu)i!}\Lambda\left(\mu+i, \frac{\mu(1+k)\gamma_0}{\bar{\gamma}}\right)\right)$$

$$f_3 = \sum_{p=0}^{\infty} \frac{\mu^{p+1}\kappa^p(1+\kappa)\Lambda\left(p+\mu-1, \mu(1+k)\gamma_0/\bar{\gamma}\right)}{e^{\mu\kappa}\bar{\gamma}\,\Gamma(p+\mu)p!}. \tag{8.42}$$

Furthermore, the case when MRC diversity is applied can be modeled by

$$\langle C \rangle_{tifr}^{MRC} = B \log_2 \left(1 + \frac{1}{\sum_{p=0}^{\infty} f_4} \right)$$

$$\times \left(1 - \sum_{i=0}^{\infty} \frac{(kL\mu)^i}{e^{L\mu k}\Gamma(i+L\mu)i!} \Lambda\left(\mu L + i, \frac{\mu L(1+k)\gamma_0}{\gamma} \right) \right)$$

$$f_4 = \sum_{p=0}^{\infty} \frac{\mu^{p+1}\kappa^p(1+\kappa)L^p\Lambda\left(p+\mu L-1,\mu L(1+k)\gamma_0/\overline{\gamma}\right)}{e^{\mu\kappa L}\overline{\gamma}\,\Gamma(p+\mu L)p!}. \qquad (8.43)$$

The convergence of infinite series expressions in (8.42) and (8.43) is rapid, since we need about 10–15 terms to be summed in order to achieve accuracy at the 5th significant digit.

8.5.2 Weibull Fading Channels

After substituting (8.7) into (8.41) we can obtain the expression for the CIFR channel capacity over Weibull fading channels when MRC diversity is applied in the following form:

$$\frac{\langle C \rangle_{tifr}^{MRC}}{B} = \log_2 \left(1 + \frac{\Xi\overline{\gamma}\Gamma(L)}{\Gamma\left(L-2/\beta,(\gamma_0/\Xi\overline{\gamma})^{\beta/2}\right)} \right) \frac{\Gamma\left(L,(\gamma_0/\Xi\overline{\gamma})^{\beta/2}\right)}{\Gamma(L)}.$$

$$(8.44)$$

8.6 Numerical Results

In order to discuss the use of diversity techniques and adaptation policies and to show the effects of various system parameters on obtained channel capacity, numerically obtained results are graphically presented.

In Figures 8.1 and 8.2, channel capacity without diversity, for the cases when κ-μ and Weibull fading affects channels, respectively, for

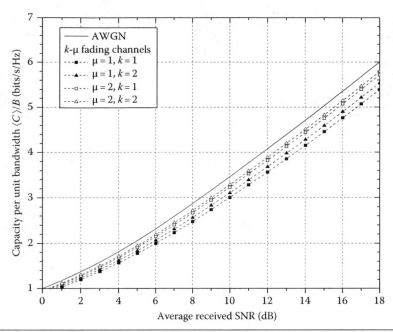

Figure 8.1 Channel capacity per unit bandwidth over k-μ fading channels and an AWGN channel versus average received SNR.

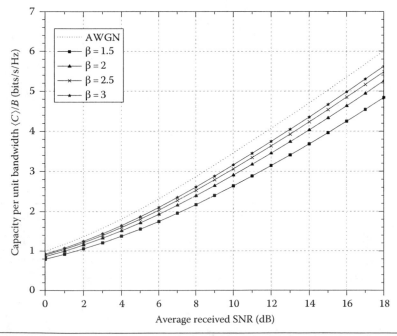

Figure 8.2 Channel capacity per unit bandwidth over Weibull fading channels and an AWGN channel versus average received SNR.

various system parameters are plotted against the average received SNR. These figures also display the capacity of an AWGN channel, C_{AWGN}, given by

$$C_{AWGN} = B \log_2(1+\gamma). \tag{8.45}$$

Considering the obtained results, with respect to $C_{AWGN} = 3.46$ dB for the average received SNR $\gamma = 10$ dB, we find that depending on fading parameters of κ-μ and Weibull distribution, channel capacity could be reduced up to 30%. From Figure 8.1, we can see that channel capacity is less reduced for the cases when fading severity parameter μ, and dominant/scattered components power ratio κ, have higher values, since for smaller κ and μ values the dynamics in the channel is larger. A similar notation can be observed from Figure 8.2 where the channel capacity is less reduced in the areas where the Weibull fading parameter β has higher values.

Figures 8.3 through 8.6 show the channel capacity per unit bandwidth as a function of $\overline{\gamma}$ for the different adaptation policies with MRC diversity over the κ-μ fading channels. It can be seen that as

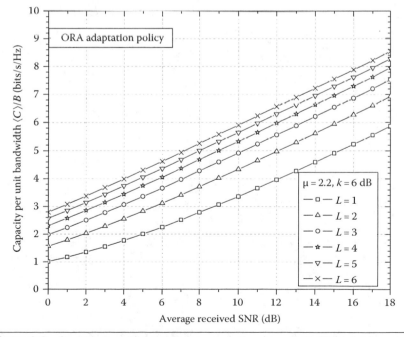

Figure 8.3   Capacity per unit bandwidth over κ-μ fading channel with ORA policy for various values of MRC diversity order.

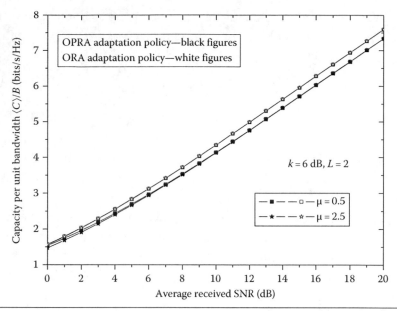

Figure 8.4   Capacity per unit bandwidth over κ-μ fading channel with different adaptation policies for dual-branch MRC diversity reception.

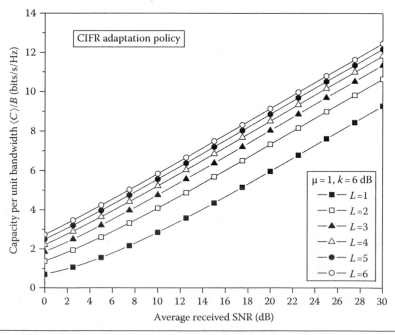

Figure 8.5   Capacity per unit bandwidth over κ-μ fading channel with CIFR policy for various values of MRC diversity order.

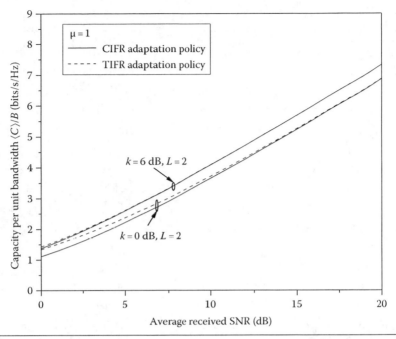

Figure 8.6 Capacity per unit bandwidth over κ-μ fading channel with MRC diversity reception for different values of fading parameter.

the number of combining branches increases the fading influence is progressively reduced, so the channel capacity improves remarkably. The greatest improvement is obtained in going from a single-branch to a two-branch combining case. However, as L increases, all capacities of the various policies converge to the capacity of an array of L independent AWGN channels, given by

$$C_{AWGN}^{MRC} = B\log_2(1 + L\gamma). \tag{8.46}$$

Figure 8.4 shows the calculated channel capacity per unit bandwidth of two different adaptation policies. From this figure we can see that the OPRA policy yields a noticeable increase in capacity over the ORA policy, and this increase also exists when the fading channel condition is changed from $\mu = 0.5$ to $\mu = 2.5$. Therefore, the ORA policy suffers a larger capacity penalty relative to the OPRA adaptation policy.

Figure 8.6 also shows the channel capacity of two adaptation policies, CIFR and TIFR, for different values of fading parameter k. The case with the TIFR policy yields a higher capacity compared to the CIFR adaptation policy for different values of k factor. The difference between these two policies is greater for small values of parameter k. Therefore, for environments where the LOS component is evidently strong, the TIFR adaptation policy is a better alternative than the CIFR adaptation policy for the proposed MRC diversity case.

Thus, in practice, it is not possible to entirely eliminate the effects of fading through space diversity since the number of diversity branches is limited. Also considering downlink (base station to mobile) implementation, we found that mobile receivers are generally constrained in size and power.

In Figure 8.7, a comparison of the channel capacity per unit bandwidth with the CIFR adaptation policy, when SC and MRC diversity techniques are applied at the reception, is shown. As expected, better performances are obtained when MRC reception over κ-μ fading channels is applied.

Figure 8.8 shows the calculated channel capacity per unit bandwidth as a function of $\overline{\gamma}$ for different adaptation policies. From this figure, we can see that the OPRA protocol yields a small increase in capacity over ORA adaptation, and this small increase in capacity

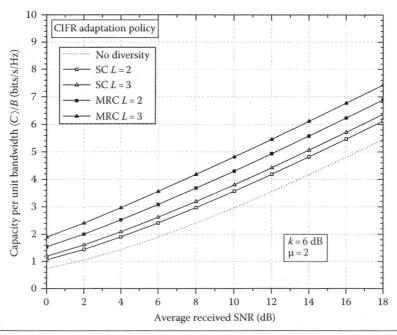

Figure 8.7   Capacity per unit bandwidth over κ-μ fading channel with CIFR policy for MRC and SC diversity reception.

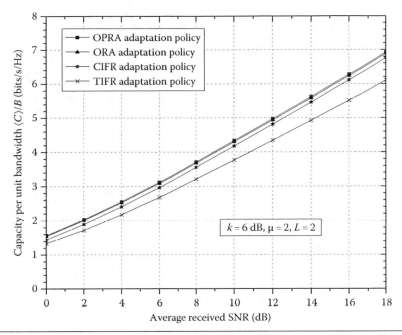

Figure 8.8   Comparison of adaptation policies over κ-μ fading channel with MRC diversity reception.

diminishes as $\overline{\gamma}$ increases. However, a greater improvement is obtained in going from complete to truncated channel inversion policy. The truncated channel inversion policy provides better diversity gain compared to complete channel inversion varying any of the parameters.

Similar results are presented considering channels affected by Weibull fading. Figures 8.9 through 8.12 show the channel capacity per unit bandwidth as a function of $\overline{\gamma}$ for the different adaptation policies with L-branch MRC diversity applied over Weibull fading channels. A comparison of adaptation policies is presented in Figure 8.13.

The nested infinite sums in (8.38) and (8.39), as can be seen from Table 8.1, for dual- and triple-branch diversity cases, converge for any value of the parameters κ, μ, and $\overline{\gamma}$. As shown in Table 8.1, the number of the terms that need to be summed to achieve a desired accuracy, depends strongly on these parameters, and it increases as these parameter values increase.

Cases when wireless channels are affected by general and nonlinear fading distributions are discussed in this chapter. The analytical study of the κ-μ fading channel capacity, for example, under the OPRA, ORA,

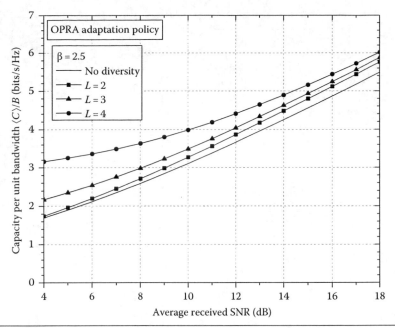

Figure 8.9   Capacity per unit bandwidth over Weibull fading channel with OPRA policy for various values of MRC diversity order.

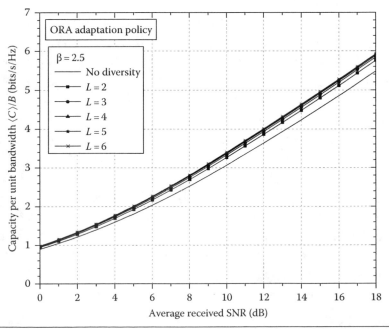

Figure 8.10   Capacity per unit bandwidth over Weibull fading channel with ORA policy for various values of MRC diversity order.

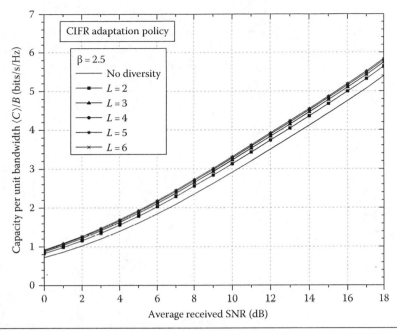

Figure 8.11   Capacity per unit bandwidth over Weibull fading channel with CIFR policy for various values of MRC diversity order.

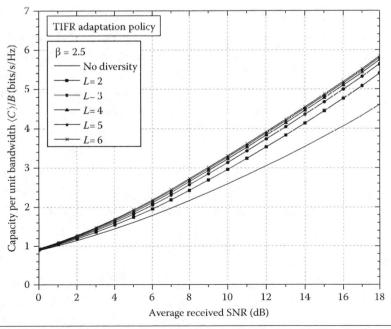

Figure 8.12   Capacity per unit bandwidth over Weibull fading channel with TIFR policy for various values of MRC diversity order.

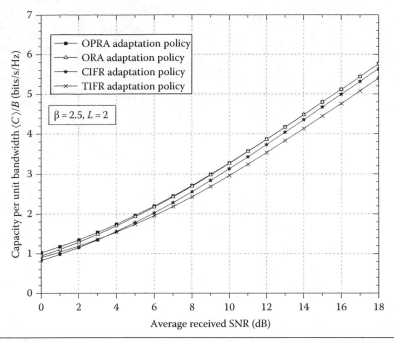

Figure 8.13   Comparison of adaptation policies over Weibull fading channel with MRC diversity reception.

CIFR, and TIFR adaptation policies and the MRC and SC diversity techniques, is performed. The main contributions are the expressions derived for the proposed adaptation policies and diversity techniques. Based on them, numerically obtained results are graphically presented to show the effects of various system parameters. Since the κ-μ model as a general physical fading model includes Rayleigh, Rician, and Nakagami-*m* fading models, as special cases, the generality and applicability of this analysis are more than obvious. The nonlinear fading scenario is discussed in a similar manner, as an analytical study of the Weibull fading channel capacity, under the OPRA, ORA, CIFR, and TIFR adaptation policies and the MRC diversity technique.

References

1. Goldsmith, A. and Varaiya, P. (1997). Capacity of fading channels with channel side information. *IEEE Transactions on Information Theory*, 43(6), 1896–1992.
2. Freeman, L. R. (2005). *Fundamentals of Telecommunications*. John Wiley & Sons, Hoboken, NJ.

3. Sampei, S., Morinaga, N., and Kamio, Y. (1995). Adaptive modulation/TDMA with a BDDFE for 2 Mbit/s multimedia wireless communication systems. *Proceedings of the IEEE Vehicular Technology Conference*, Chicago, Illinois, pp. 311–315.

4. Lee, W. C. Y. (2001). *Mobile Communications Engineering*. McGraw-Hill, London, U.K.

5. Ibnkahla, M. (2000). *Signal Processing for Mobile Communications*. CRC Press, Boca Raton, FL.

6. Brennan, D. (1959). Linear diversity combining techniques. *Proceedings of Institute of Radio Engineers*, Ottawa, Ontario, Canada. Vol. 47, pp. 1075–1102.

7. Alouini, M. S. and Goldsmith, A. (1999). Capacity of Rayleigh fading channels under different adaptive transmission and diversity-combining techniques. *IEEE Transactions on Vehicular Technology*, 48(4), 1165–1181.

8. Simon, M. K. and Alouini, M. S. (2005). *Digital Communications over Fading Channels*, 2nd edn. Wiley, New York.

9. Shao, J., Alouini, M., and Goldsmith A. (1999). Impact of fading correlation and unequal branch gains on the capacity of diversity systems. *Proceedings of the IEEE Vehicular Technology Conference*, Amsterdam, the Netherlands, pp. 2159–2163.

10. Bhaskar, V. (2009). Capacity evaluation for equal gain diversity schemes over Rayleigh fading channels. *International Journal of Electronics and Communications (AEUE)*, 63(4), 235–240.

11. Anastasov, J., Panic, S., Stefanovic M., and Milenkovic, V. Capacity of correlative Nakagami-m fading channels under adaptive transmission and maximal-ratio combining diversity technique. *Journal of Communications Technology and Electronics* (accepted for publication).

12. Subadar, R. and Sahu, P. (2011). Channel capacity of adaptive transmission schemes using equal gain combining receiver over Hoyt fading channels. *Proceedings of National Conference on Communications*, Bangalore, India, pp. 1–5.

13. Sagias, N. C. (2006). Capacity of dual-branch selection diversity receivers in correlative Weibull fading. *European Transactions on Telecommunications*, 16(1), 37–43.

14. Yacoub, M. D. (2007). The κ-μ distribution and the η-μ distribution. *IEEE Antennas and Propagation Magazine*, 49(1), 68–81.

15. Milisic, M., Hamza, M., and Hadzialic M. (2009). BEP/SEP and outage performance analysis of L-branch maximal-ratio combiner for κ-μ fading. *International Journal of Digital Multimedia Broadcasting*, 2009, Article ID 573404, 1–8.

16. Gradshteyn, I. and Ryzhik, I. (1980). *Tables of Integrals, Series, and Products*. Academic Press, New York.

17. Sagias, N. C. et al. (2004). Channel capacity and second order statistics in Weibull fading. *IEEE Communications Letters*, 8(6), 377–379.

18. Filho, J. and Yacoub, M. D. (2006). Simple precise approximations to Weibull sums. *IEEE Communications Letters*, 10(8), 614–616.

19. Sagias, N. C. and Mathiopoulos, P. T. (2005). Switched diversity receivers over generalized gamma fading channels. *IEEE Communications Letters*, 9(10), 871–873.

20. Abramowitz, M. and Stegun, I. (1970). *Handbook of Mathematical Functions*. Dover Publications, Inc., Mineola, NY.

21 Academik, V. and Marichev, O. (1990). The algorithm for calculating integrals of hypergeometric type functions and its realization in REDUCE system. *Proceedings of International Conference on Symbolic and Algebraic Computation*, pp. 212–224.

22. Wolfram Research, Inc. http:/functions.wolfram.com (accessed June 2012).

23. Prudnikov, A., Brychkov, Y., and Marichev, O. (1990). *Integral and Series: Volume 3, More Special Functions*. CRC Press Inc., Boca Raton, FL.

Appendix A

A.1 Random Variables

Random variable (RV) defines a mapping of the sample space Ω into the set of real numbers [1]. According to the definition, RV takes a value from the set of real numbers and is the function with a set domain of Ω.

There are various conventional ways for denoting RVs. RV is most commonly denoted with capital letters X, Y, Z, ... although small letters from the Greek alphabet ξ, η, ζ could also be used.

Classification of RVs is based on the fact whether sample space is finite (countable) or event contains an innumerable set of elementary events.

In the first case, we deal with *discrete RVs*. Then the index i covers all possible events and stands

$$\sum_i P(\xi = x_i) = 1. \tag{A.1}$$

In the second case, we deal with *continuous RVs*.

A.2 Cumulative Distribution Function and Probability Density Function

Discrete RV ξ is defined with a set of values $x_i(i = 1, 2, ...)$ and corresponding probabilities:

$$P(\xi = x_i) = P_\xi(x_i), \quad \sum_i P_\xi(x_i) = 1 \qquad (A.2)$$

These two sets (values of RV and corresponding probabilities) are also called distribution law. However, continuous RV (because of unaccountability of probably results) requires a special approach. Since continuous variables could take values that vary continuously within each observed interval, an uncountable infinite number of individual outcomes arises, with each of them having zero probability. Therefore, there is a need to indicate the density of probability in a small interval around the observed value. This indication is achieved by using the probability density function (PDF) for continuous RVs.

If ξ is an RV, then the cumulative distribution function (CDF) of this RV is real function, defined as

$$F_\xi(x) = P(\xi \leq x), \quad -\infty < x < \infty. \qquad (A.3)$$

Some important properties of CDF are as follows:

1. $0 \leq F_\xi(x) \leq 1 (\forall x \in \mathfrak{R})$.
2. $F_\xi(x)$ is a *monotonically* nondecreasing function.
3. $F_\xi(x)$ is the right continuous function ($\forall x \in \mathfrak{R}$).
4. $F_\xi(x)$ has a limiting value from the left $-F_\xi(x_-)$ ($\forall x \in \mathfrak{R}$).
5. $F_\xi(-\infty) = 0$, $F_\xi(+\infty) = 1$.
6. $P(x_1 < \xi \leq x_2) = F_\xi(x_2) - F_\xi(x_1)$.

Analogously to characteristic (6) $P(x_1 < \xi \leq x_2) = F_\xi(x_{2-}) - F_\xi(x_1)$, $P(x_1 \leq \xi < x_2) = F_\xi(x_2) - F_\xi(x_{1-})$, $P(x_1 \leq \xi < x_2) = F_\xi(x_{2-}) - F_\xi(x_{1-})$, $P(\xi = x_1) = F_\xi(x_1) - F_\xi(x_{1-})$, $P(\xi \leq x_1) = F_\xi(x_{1-})$, $P(\xi > x_1) = 1 - F_\xi(x_1)$. Those CDF relations stand, whether the RV takes some discrete values or values from some intervals.

The PDF could be introduced with the following definition:

Let $F_\xi(x)$ be the CDF of an RV ξ. If there is a nonnegative function $f(x)$, defined on $\mathfrak{R}$, such that for ($\forall x \in \mathfrak{R}$) stands

$$F_\xi(x) = \int_{-\infty}^{x} f_\xi(x) dx, \qquad (A.4)$$

then it is said that continuous RV, ξ, has a corresponding PDF $w_\xi(x)$.

If the CDF is not only continuous but also differentiable with a continuous derivation everywhere, except eventually, in final or countable set of points, then the PDF can be expressed as a derivation of CDF, that is,

$$f_\xi(x) = \frac{d}{dx} F_\xi(x). \tag{A.5}$$

Starting from the definition of the first derivation, the PDF can also be written as

$$P(x - dx < \xi \le x) = f_\xi(x)dx. \tag{A.6}$$

Some properties of PDF are as follows:

1. For continuous RV, it stands

$$P(x_1 < \xi < x_2) = \int_{x_1}^{x_2} f_\xi(x)dx.$$

   Besides, every sign $<$ could be changed with $\le$.
2. For continuous RV, it stands

$$P(\xi = x_1) = 0.$$

3. Generally, it stands

$$\int_{-\infty}^{\infty} w_\xi(x)\,dx = 1.$$

A.3 Random Vectors and Independency of Random Variables

When a few RVs are observed in the sample space, then an ordered set of these variables is called a *random vector* or *multidimensional random variable*. In the two-dimensional case for every elementary event, there corresponds an element from space $\Re \times \Re = \Re^2$. In addition, a complex RV can be defined as $\zeta = \xi + j\eta$, where ξ and η are RVs. The density of a complex random variable $w_\zeta(y) = w_{\xi\eta}(x, z)$ is a real-valued function [2]. We will analyze only two-dimensional RVs case—vector (ξ, η), while multidimensional concepts could be introduced analogously.

The joint CDF of random vector (ξ, η) is defined as

$$F_{\xi\eta}(x, y) = P(\xi \leq x, \eta \leq y), \quad -\infty < x, y < \infty. \tag{A.7}$$

Some properties of this function are as follows:

1. $0 \leq F_{\xi\eta}(x, y) \leq 1$.
2. $F_{\xi\eta}(x, y)$ is not a monotonically decreasing function for both variables.
3. $F_{\xi\eta}(x, y)$ is a right continuous function for both variables.
4. $F_{\xi\eta}(-\infty, -\infty) = 0$, $F_{\xi\eta}(\infty, \infty) = 1$.
5. $P(x_1 < \xi \leq x_2, x_1 < \eta \leq y_2) = F_{\xi\eta}(x_2 y_2) - F_{\xi\eta}(x_2 y_1) - F_{\xi\eta}(x_1 y_2) + F_{\xi\eta}(x_1 y_1)$.

Completely analogous can be defined by the joint PDF (JPDF), $f_{\xi\eta}(x, y)$:

$$F_{\xi\eta}(x, y) = \int\limits_{-\infty}^{x} \int\limits_{-\infty}^{y} f_{\xi\eta}(x, y)\,dxdy. \tag{A.8}$$

Under the analogue conditions, the JPDF could be obtained by using correspondent differentiation

$$f_{\xi\eta}(x, y) = \frac{\partial^2}{\partial x \partial y} F_{\xi\eta}(x, y). \tag{A.9}$$

Marginal CDF could be found from the joint CDF as follows:

$$F_{\xi}(x) = F_{\xi\eta}(x, \infty) = \int\limits_{-\infty}^{\infty} dy \int\limits_{-\infty}^{x} f_{\xi\eta}(x, y) dx,$$

$$F_{\eta}(y) = F_{\xi\eta}(\infty, y) = \int\limits_{-\infty}^{\infty} dx \int\limits_{-\infty}^{x} f_{\xi\eta}(x, y) dy. \tag{A.10}$$

By differentiating the previous expressions, appropriate expressions for *marginal PDFs* could be obtained as

$$f_{\xi}(x) = \int\limits_{-\infty}^{\infty} f_{\xi\eta}(x, y) dy, \quad f_{\xi}(y) = \int\limits_{-\infty}^{\infty} f_{\xi\eta}(x, y)\,dx. \tag{A.11}$$

that is, by integrating the JPDF, the corresponding RV is eliminated [3]. In a similar manner, conditional CDF and conditional PDF could be introduced.

When discrete RVs are considered, then *conditional probabilities* are defined with

$$P\left(\frac{y_j}{x_i}\right) = \frac{P(x_i, y_j)}{P(x_i)}, \quad P(x_i) \neq 0. \tag{A.12}$$

For continuous RVs, *the conditional PDF* is given by

$$f_{\eta/\xi}\left(\frac{y}{x}\right) = \frac{f_{\xi\eta}(x, y)}{f_\xi(x)}, \quad f_\xi(x) \neq 0. \tag{A.13}$$

Also,

$$\int_{-\infty}^{\infty} f_{\eta/\xi}\left(\frac{y}{x}\right) dy = 1. \tag{A.14}$$

Generally, two RVs are independent if and only if

$$F_{\xi\eta}(x, y) = F_\xi(x) F_\eta(y). \tag{A.15}$$

The corresponding relation for discrete RVs is

$$P(x_i, y_j) = P(x_i) P(y_j), \forall i, j \tag{A.16}$$

while an equivalent relation for the continuous RVs is

$$f_{\xi\eta}(x, y) = f_\xi(x) f_\eta(y). \tag{A.17}$$

Therefore, the independent RVs JCDF equals the joint product of marginal CDFs, and if RVs are continuous then the JPDF can be written as a product of marginal PDFs [4].

A.4 RV Functions and Transformation of PDFs

As expected, the function of RV is also an RV with its own distribution and density. The same is the case for the multivariable function. Starting from the transformation law function itself and the known PDF (or JPDF), the PDF of the corresponding function should be determined.

A.4.1 Transformations of Single RV

Let ξ be an RV with its PDF $w_\xi(x)$. Let us also define a new RV $\eta = f(\xi)$, that is, $y = f(x)$. We have to determine the new PDF $w_\eta(y)$. Let us consider the case where the correspondence between variables ξ and η is 1:1, that is, a single value of η corresponds to a single value ξ (and vice versa). Based on the definition of the PDF, it can be written

$$P\left(x < \xi < x + dx\right) = P\left(y < \eta < y + dy\right) \Rightarrow w_\xi(x)dx = w_\eta(y)dy. \quad \text{(A.18)}$$

Since the continuity of an RV is presumed, instead of a sign $\leq$ sign $<$ is used here. Since the resulting PDF should be a function of variable y, we should find the inverse function of $x = f^{-1}(y) = g(y)$ and substitute it in the result obtained:

$$w_\eta(y) = \frac{w_\xi(x)}{|dy/dx|}\bigg|_{x=g(y)} = \frac{w_\xi(x)}{|f'(x)|}\bigg|_{x=g(y)}. \quad \text{(A.19)}$$

An absolute value sign has been introduced because of the requirement that the PDF must be nonnegative. For the cases where the correspondence between RVs is n:1, particular functional relationships must be taken into account.

A.4.2 Transformations of Multiple Random Variables

A very important scenario is when the resulting RV equals the sum of two independent RVs, that is, $\zeta = \xi + \eta$ ($z = x + y$) (this is mapping $\mathfrak{R}^2$ into $\mathfrak{R}$). Then, the resulting PDF is obtained as a *convolution of densities*, that is,

$$w_\zeta(z) = \int_{-\infty}^{\infty} w_\xi(x)w_\eta(z-x)\,dx = \int_{-\infty}^{\infty} w_\eta(y)w_\xi(z-y)\,dy = w_\xi(z) * w_\eta(z).$$

$$\text{(A.20)}$$

This result could also be expanded on the sum of a higher number of independent random variables.

The previous result can be derived as a special case of general consideration: the transformation of the sum of n RVs (of random vector $\Xi(\xi_1, \xi_2, \dots, \xi_n)$), whose JPDF is already known, into a novel set of

n RVs $E(\eta_1, \eta_2,\ldots, \eta_n)$. Let the novel RVs be unambiguous functions of the observed RVs with continuous partial derivatives:

$$\begin{aligned}
\eta_1 &= f_1(\xi_1, \xi_2,\ldots, \xi_n) \\
\eta_2 &= f_2(\xi_1, \xi_2,\ldots, \xi_n) \\
\eta_n &= f_n(\xi_1, \xi_2,\ldots, \xi_n).
\end{aligned} \tag{A.21}$$

Let us also assume that the original RVs could be expressed as unambiguous functions of the newly introduced RVs, that is,

$$\begin{aligned}
\xi_1 &= g_1(\eta_1, \eta_2,\ldots, \eta_n) \\
\xi_2 &= g_2(\eta_1, \eta_2,\ldots, \eta_n) \\
\xi_n &= g_n(\eta_1, \eta_2,\ldots, \eta_n).
\end{aligned} \tag{A.22}$$

Then, the novel JPDF can be given in the form of

$$w_E\left(y_1, y_2,\ldots, y_n\right) = w_\Xi(x_1, x_2,\ldots, x_n)|J| \tag{A.23}$$

where $|J|$ stands for the absolute value of Jacobian transformation:

$$J(x_1,\ldots, x_n) = \begin{vmatrix} \dfrac{\partial g_1}{\partial x_1} & \cdots & \dfrac{\partial g_1}{\partial x_n} \\ \cdots & \cdots & \cdots \\ \dfrac{\partial g_n}{\partial x_1} & \cdots & \dfrac{\partial g_n}{\partial x_n} \end{vmatrix}, \tag{A.24}$$

and all of the $x_i(i=1,2,\ldots, n)$ values should be changed with $y_j(j=1,2,\ldots, n)$ according to the previous set of equations.

If the number of newly introduced RVs k is smaller than the number of RVs that should be transformed, $(k < n)$, then additional variables should be used for supplementation, that is,

$$y_{k+1} = x_{k+1},\ldots, y_n = x_n \tag{A.25}$$

and this supplementation should be followed by integration to eliminate these redundant variables, that is,

$$w_{E_k}\left(y_1, y_2,\ldots, y_k\right) = \int_{-\infty}^{\infty} \cdots \int_{-\infty}^{\infty} w_{E_k}\left(y_1, y_2,\ldots, y_n\right) dy_{k+1} dy_{k+2} \cdots dy_n, \tag{A.26}$$

with $(n - k)$ integrations.

If the number of newly introduced RVs k is larger than the number of RVs that should be transformed, $(k > n)$, then variables $y_{n+1}, \ldots, y_k$ which has a total of $(n - k)$, should be expressed through the variables $y_1, \ldots, y_n$.

A.5 Numerical Characteristics of RVs

Random variables are described by their PDFs and CDFs. Hence, it is also useful to observe some numerical characteristics that describe some of their properties.

A.5.1 Mathematical Expectation (Average Value)

Mathematical expectation, also known as statistical average or mean value, of discrete RV, ξ, with distribution $P_\xi(x_i)$, $(i = 1, 2, \ldots)$, is defined as

$$E[\xi] = \overline{\xi} = m_\xi = \sum_i x_i P_\xi(x_i). \tag{A.27}$$

Similarly, for the case of continuous RV with PDF $w_\xi(x)$, it is defined as

$$E[\xi] = \int_{-\infty}^{\infty} x\, w_\xi(x)\, dx. \tag{A.28}$$

If series or integrals do not absolutely converge, or if they diverge, then mathematical expectation does not exist. The value $E[\xi]$ could be outside the set of RV.

The average value of the function of random variable $\eta = f(\xi)$, that is, $y = f(x)$, which itself is an RV, is given as

$$E[\eta] = \int_{-\infty}^{\infty} y w_\eta(y)\, dy = \int_{-\infty}^{\infty} f(x) w_\xi(x)\, dx, \tag{A.29}$$

while for discrete RV, it is given as

$$\overline{f(\xi)} = \sum_i f(x_i) P_\xi(x_i). \tag{A.30}$$

For RVs with JPDF $w_{\xi\eta}(x, y)$, the average value of function $f(\xi, \eta)$ is given as

$$\overline{f(\xi, \eta)} = \int\limits_{-\infty}^{\infty}\int\limits_{-\infty}^{\infty} f(x, y) w_{\xi\eta}(x, y)\, dx dy, \tag{A.31}$$

while for discrete RVs with distribution $P_{\xi\eta}(x_i, y_j)$

$$\overline{f(\xi, \eta)} = \sum_i \sum_j f(x_i, y_j) P_{\xi\eta}(x_i, y_j). \tag{A.32}$$

The average value of function can be defined completely analogue also for multidimensional RVs. It should be noted that in the process of determining the average function of RV, it is not necessary to determine the PDF of a transformed variable (or variables) [5].

The basic characteristic of this operator is its linearity, namely, the operator of averaging is linear. Hence, the following properties arise (ξ and η are random variables while a, b, and c are real numbers):

1. $E[c] = c$, that is, mean value of the constant is the constant.
2. $E[a\xi + b\eta] = aE[\xi] + bE[\eta]$.
3. If the RVs are statistically independent, then

$$\overline{(\xi \cdot \eta)} = \overline{\xi} \cdot \overline{\eta}.$$

In addition, a *conditional average value* (over conditional probability) can be defined. For discrete RV, ξ, the conditional average value after event A is defined as

$$E\left[\frac{\xi}{A}\right] = \sum_i x_i P\left(\xi = \frac{x_i}{A}\right), \tag{A.33}$$

while for continuous RV, it is defined as

$$E\left[\frac{\xi}{A}\right] = \int\limits_{-\infty}^{\infty} x w_\xi\left(\frac{x}{A}\right) dx. \tag{A.34}$$

Mathematical expectation is often used as a numerical property connected in some manner with the center of the PDF.

A.5.2 Variance, Correlation, and Covariance

As it will be seen, variance and covariance are special cases of moments of RVs. Because of their importance, they have a major role. The average value itself does not provide any information about "dissipation" (dispersion) or concentration of RV near its average value. Corresponding numerical characteristics, which provide this kind of information, are variance (average square deviation from its mean value) and standard deviation (square root of variance).

Let ξ be a continuous RV whose PDF is $w_\xi(x)$ and with average value of $E_\xi = m_\xi$. Then, its corresponding *variance* is defined as

$$\sigma_\xi^2 = E\left[\left(x - m_\xi\right)^2\right] = \int_{-\infty}^{\infty} \left(x - m_\xi\right)^2 w_\xi(x)\,dx, \qquad (A.35)$$

while for discrete RV, with distribution $P_\xi(x_i)$, the variance is defined as

$$\sigma_\xi^2 = \sum_i \left(x_i - m_\xi\right)^2 P_\xi(x_i). \qquad (A.36)$$

Notation $Var[\xi]$ also stands for the variance, while notation σ_ξ denotes the *standard deviation*.

The most important properties of variance are (a—real constant) as follows:

1. $Var[a] = 0$.
2. $E\left[\left(\xi - m_\xi\right)^2\right] = E\left[\xi^2\right] - E^2[\xi] = E\left[\xi^2\right] - m_\xi^2$　or　otherwise stated $\overline{(\xi - \overline{\xi})^2} = \overline{\xi^2} - \overline{\xi}^2$.
3. $Var[\xi + a] = Var[\xi]$.
4. $Var[a\xi] = a^2 Var[\xi]$.
5. If ξ and η are statistically independent (with finite variances), $Var[\xi + \eta] = Var[\xi] + Var[\eta]$.

Variance has its own physical interpretation, depending on the field of application. In electrical engineering, especially in telecommunications, it corresponds to the average of the observed signal strength [6].

Correlation of two continuous RVs ξ and η with the JPDF $w_{\xi\eta}(x, y)$ by definition is equal to the average value of their product

$$R_{\xi\eta} = \overline{\xi} \cdot \overline{\eta} = \int\limits_{-\infty}^{\infty} \int\limits_{-\infty}^{\infty} xy w_{\xi\eta}(x, y)\, dx\, dy. \tag{A.37}$$

If the RVs are discrete with joint distribution $P_{\xi\eta}(x_i, y_j)$, then the corresponding correlation follows:

$$R_{\xi\eta} = \sum_i \sum_j x_i y_j P_{\xi\eta}(x_i, y_j). \tag{A.38}$$

If the RVs are statistically independent, then the average value of their product (i.e., their correlation) equals the product of their mean values, namely $\overline{(\xi \cdot \eta)} = \overline{\xi} \cdot \overline{\eta}$. It should be emphasized again that the reverse may not be worth, that is, the previous relation does not necessarily imply the statistical independence of the observed variables [7].

The *covariance* of two RVs is defined as the average value of their products, where they are seen centered relative to their mean values, that is,

$$Cov(\xi, \eta) = C_{\xi\eta} = \overline{(\xi - m_\xi)(\eta - m_\eta)} = \overline{\xi\eta} - m_\xi m_\eta = R_{\xi\eta} - m_\xi m_\eta. \tag{A.39}$$

If the RVs are statistically independent, then their covariance is zero (reverse may not be worth). Therefore, if RVs are statistically independent, their correlation is equal to the product of their mean values, while the covariance is zero. Of course, if at least one of the RVs has mean zero, then the correlation between independent variables is also zero [8].

The relation between covariance and variance is defined as

$$Var[\xi + \eta] = Var[\xi] + Var[n] + 2 \cdot Cov(\xi, \eta). \tag{A.40}$$

In addition to the previously mentioned statistical properties of the covariance of two independent RVs, there are some additional covariance properties of interest such as

1. $Cov(\xi, \xi) = Var(\xi)$
2. $Cov(a\xi, b\eta) = ab\, Cov(\xi, \eta)$
3. $Cov(\xi + \eta, \zeta) = Cov(\xi, \zeta) + Cov(\eta, \zeta)$
4. $Cov(a + \xi, b + \eta) = Cov(\xi, \eta)$

(a and b are real constants)

The last property implies that the RVs ξ and η and centered RVs $\xi - m_\xi$ and $\eta - m_\eta$ have the same covariance.

If the correlation between two RVs is zero, then they are *orthogonal* variables. If two RVs have zero covariance, then they are uncorrelated variables. This terminology comes from the definition of the *correlation coefficient*:

$$\rho(\xi, \eta) = \frac{C_{\xi\eta}}{\sigma_\xi \sigma_\eta}. \tag{A.41}$$

This coefficient presents normalized covariance and equals zero when RVs are independent. On the contrary, if the correlation coefficient equals zero, variables are uncorrelated, but they are not necessarily independent.

A.5.3 Correlation and Covariance Matrix

Let us observe random vector $\Xi(\xi_1, \xi_2, \ldots, \xi_n)$. By definition, *correlation matrix* is given as

$$R = E\left[\Xi^T \Xi^* \right] = \begin{bmatrix} R_{11} & \cdots & R_{1n} \\ \vdots & \ddots & \vdots \\ R_{n1} & \cdots & R_{nn} \end{bmatrix}, \tag{A.42}$$

where $(\cdot)^T$ indicates the transposition operator, and $(\cdot)^*$ the conjugation operator (for a complex random vector). Correlations between the corresponding vector components are elements of a matrix. If the matrix is singular, then variables are linearly dependable, that is,

$$a_1 \xi_1 + \cdots + a_n \xi_n = 0. \tag{A.43}$$

For a diagonal matrix, the determinant is of the form

$$\Delta_n = R_{11} R_{22} \ldots R_{nn}. \tag{A.44}$$

The *Covariance matrix* is defined as:

$$C_n = \begin{bmatrix} C_{11} & \cdots & C_{1n} \\ \vdots & \ddots & \vdots \\ C_{n1} & \cdots & C_{nn} \end{bmatrix}, \tag{A.45}$$

with the elements given as

$$C_{ij} = R_{ij} - m_i m_j^* = C_{ji}^*, \quad R_{ij} = E\xi_i \xi_j = C_{ji}^*, \tag{A.46}$$

or written in matrix form:

$$C_n = R_n - m_\Xi^T m_\Xi^*, \tag{A.47}$$

with m_Ξ being the vector of mean values.

If the covariance matrix is diagonal, RVs are *uncorrelated*. In that case, a novel RV is defined as the sum of RVs:

$$\eta = \xi_1 + \cdots + \xi_n, \tag{A.48}$$

has variance equal to the sum of variances of its components:

$$\sigma_\eta^2 = \sum_{i=1}^{n} \sigma_{\xi_i}^2. \tag{A.49}$$

The covariance matrix of a random vector is a symmetrical positive-definitive matrix. This means that for arbitrary vector $a[a_1, a_2, ..., a_n]$ stands

$$a C a^T \geq 0. \tag{A.50}$$

In scalar form, the previous relation could be expressed as

$$\sum_{i=1}^{n}\sum_{j=1}^{n} a_i C_{ij} a_j \geq 0. \tag{A.51}$$

If A is also an n-order nonsingular quadrant matrix, for random vector Ξ, covariance matrix $E = A\Xi$ could be determined according to

$$C_E = A C_\Xi A^T. \tag{A.52}$$

A.5.4 Moments

It has been previously indicated that mean value, variance, and correlation are only special cases of general forms-moments. There are several types of moments.

Moment (*initial*) is defined as the mean value of the RV powered at the nth degree. For discrete RV, it is defined as

$$m_n = E\left[\xi^n\right] = \overline{\xi^n} = \sum_i x_i^n P_\xi(x_i), \tag{A.53}$$

while for continuous RV, it is defined as

$$m_n = \int_{-\infty}^{\infty} x^n w_\xi(x)\, dx. \tag{A.54}$$

The mean value itself is the first moment, as it stands:

$$m_1 = \overline{\xi} = m_\xi = m. \tag{A.55}$$

Absolute moments are obtained by averaging corresponding absolute values:

$$E\left[|\xi|^n\right]. \tag{A.56}$$

Central moments are obtained as moments of RVs with regard to their mean values, that is,

$$\mu_n = \overline{(\xi - \overline{\xi})^n} = \int_{-\infty}^{\infty} (x - m_\xi) w_\xi(x)\, dx. \tag{A.57}$$

Accordingly, variance corresponds to the second central moment of RV ($\mu_2 = \sigma_\xi^2$).

When observing two RVs ξ and η, joint moments could be defined. The joint moment of order ($c + k$) for continuous RVs is defined as

$$m_{nk} = E\left[\xi^n \eta^k\right] = \overline{\xi^n \eta^k} = \int_{-\infty}^{\infty}\int_{-\infty}^{\infty} x^n y^k w_{\xi\eta}(x, y)\, dx\, dy, \tag{A.58}$$

while for a discrete RV, it is defined as

$$m_{nk} = \sum_i \sum_j x_i^n y_j^k P_{\xi\eta}(x_i y_j). \tag{A.59}$$

Similarly, *joint central moments* could be obtained according to

$$\mu_{nk} = \overline{(\xi - m_\xi)^n (\eta - m_\eta)^k}. \tag{A.60}$$

A.6 Characteristic Function and Moment Generating Function

The characteristic function (CHF) of continuous RV is obtained as the Fourier transform of its PDF:

$$W_\xi(j\Omega) = \int_{-\infty}^{\infty} w_\xi(x) e^{j\Omega x} dx, \tag{A.61}$$

but it should be noted that the only difference is that there is no minus sign in the exponent. For a discrete RV stands:

$$W_\xi(j\Omega) = \sum_i w_\xi(x_i) e^{j\Omega x_i}. \tag{A.62}$$

In general case, the CHF can be defined as a mean value of function $e^{j\Omega x}$, that is, as

$$W_\xi(j\Omega) = E\left[e^{j\Omega x}\right]. \tag{A.63}$$

The CHF always exists and the PDF could be unambiguously determined by its CHF (and vice versa). The PDF could be obtained by applying the inverse Fourier transform (with the minus sign in the exponent) as

$$w_\xi(j\Omega) = \frac{1}{2\pi} \int_{-\infty}^{\infty} W_\xi(j\Omega) e^{-j\Omega x} d\Omega. \tag{A.64}$$

It can be shown that

$$\left|W_\xi(j\Omega)\right| \leq W_\xi(0) = 1. \tag{A.65}$$

By differing the CHF, moments of RVs (if they exist) could be determined as

$$\overline{\xi^n} = \frac{1}{j^n} \frac{d^n W_\xi(j\Omega)}{(d\Omega)^n}\bigg|_{\Omega=0} \tag{A.66}$$

Moment generating function (*MGF*) is directly related to the CHF according to

$$W_\xi(s) = \int_{-\infty}^{\infty} w_\xi(x) e^{sx} dx = E\left[e^{sx} \right] = \sum_{i=0}^{\infty} \frac{\overline{\xi^i}}{i!} s^i . \qquad (A.67)$$

It could be obtained from CHF by substituting the real variable s instead of $j\Omega$ ("real Fourier's transform"). This makes sense only if all moments are finite and if the series converges absolutely around $s = 0$.

A.7 Central Limit Theorem

Central limit theorem (*CLT*) deals with the more general problem of asymptotic distribution of the sum of independent RVs. One of the CLT formulations is as follows:

Let ξ_1, ξ_2,..., ξ_n be n independent and identically distributed RVs with a finite mean value (m) and finite variance (σ^2). Then, normalized RV

$$\eta = \frac{\xi_1 + \xi_2 + \cdots + \xi_n - n \cdot m}{\sigma\sqrt{2}} \qquad (A.68)$$

converges in the PDF toward a suitable normalized Gaussian (Normal) distribution $N(m, \sigma^2)$, that is,

$$\lim_{n \to \infty} P\left[\eta \le x \right] = \frac{1}{2 \cdot \pi} \int_{-\infty}^{x} e^{-u^2/2} du. \qquad (A.69)$$

References

1. Drajic, D. (2003). *An Introduction to Statistical Theory of Telecommunications*. Academic Science, Belgrade, Serbia.
2. Feller, W. (1966). An *Introduction to Probability Theory and Its Applications I, II*. Wiley, New York.
3. Gnedenko, B. (1969). *The Theory of Probability*. Mir Publishers, Moscow, Russia.
4. Papoulis, A. (1984). *Probability, Random Variables and Stochastic Processes*, 2nd edn. McGraw-Hill, New York.

5. Thomas, J. B. (1971). *An Introduction to Applied Probability and Random Processes.* Wiley, New York.
6. Prolakis, J. G. (1995). *Digital Communications*, 3rd edn. McGraw-Hill, New York.
7. Gardner, W. A. (1986). *Introduction to Random Processes.* MacMillan Publishing Company, London, U.K.
8. Cooper, G. R. and McGillem, C. D. (1999). *Probabilistic Methods of Signal and System Analysis*, 3rd edn. Oxford University Press, Oxford, U.K.

Index

A

ABER, *see* Alternatively average bit
error rate (ABER)
Additive white Gaussian noise
(AWGN), 5
AFD, *see* Average fade duration
(AFD)
α-η-μ fading channels, 72–75
α-μ fading channels
multichannel receiver
SC over constantly correlated
α-μ fading channels,
98–108
SC over exponentially
correlated α-μ fading
channels, 108–119
SC over generally correlated
α-μ fading channels,
120–122
SC with uncorrelated
branches, 92–98
SSC with correlated branches,
84–93
SSC with uncorrelated
branches, 81–84

multipath fading, 11–13
multivariate correlative
constant correlation model,
30–32
exponential correlation model,
28–30
general correlation model,
32–33
selection combining (SC)
over constantly correlated,
98–108
over exponentially correlated,
108–119
over generally correlated,
120–122
with uncorrelated branches,
92–98
single channel receiver,
performance analysis
ASEP, 63
signal-to-interference ratio,
59–62
Alternatively average bit error rate
(ABER), 41
Alternatively average symbol error
rate, 41–45